Project Management & Agribusiness Entrepreneurship Development

NIPA® GENX ELECTRONIC RESOURCES & SOLUTIONS P. LTD.
New Delhi-110 034

About the Authors

Dr. Dilip Rasiklal Vahoniya is an Assistant Professor & Head, Department of Agri Entrepreneurship and Project Management, International Agribusiness Management Institute, Anand Agricultural University, Anand, Gujarat. He has specialization in Marketing Management, Digital marketing, Service Marketing, Principles of Management & Organizational Behavior, Project Management and Entrepreneurship Development, Advertising and Brand Management, Value Chain Management in Agribusiness, Advanced Entrepreneurship Development, Sales and Distribution Management, and Food Retail Management subjects. He also has experiences of total 15 years out of which, 13 years' experience in PG teaching of MBA- Agribusiness Management students and so far guided 53 PG students, one Ph. D student, delivered total 22 lectures in different universities and ICAR - NET centre. He has already Published 45 research papers in NAAS, WOS, and UGC care listed journals, also published three books and 11 book chapters. He has participated in 16 different national and international seminar & conferences. He has been handling a project of Government of Gujarat since 2016-17 to till date as a Co-PI and also successfully completed the ICAR World Bank Project as a Co-PI. He has 302.1 research interest score, 100 citations, and 4 h-index on Research Gate.

Dr. Prakashkumar Hasmukhbhai Patel, Doctorate in e-Procurement Practices in Management from Gujarat Technological University is currently associated with the Department of Commerce and Business Management, Faculty of Commerce, The Maharaja Sayajirao University of Baroda, Vadodara, Gujarat in the capacity of Assistant Professor. He's an expert in Operation management, Supply Chain Management, Marketing Management, Management Information systems, and Research. He also worked as a Consultant in a few companies in the field of Human Resources and Production & Operations Management. His research interests range from Business Ethics, E-procurement, Information Technology, e-commerce to Green Internet of Things (IoT) and Machine Learning (ML). Not only he has several publications to his credit but he has also been a valuable resource person invited as a resource person for various workshops/ seminars/ Faculty development programs in several institutions. Dr. Prakash has received many prestigious awards namely 'Teaching – Training Quality Excellence Award' based on L.E.A.D.E.R.S.H.I.P. Model for Quality excellence, 'Best Research Paper Award' at One Day

International Conference, 'Teaching Excellence Award 2018' at 6th Regional Conference on Quality & Accreditation of Higher Education in India: Policy & Best Practices, etc.

Dr. Alvirakumari Ashokbhai Rajwadi has Ph.D. (Agri.) in subject of Agricultural Economics and Awarded with NFST Fellowship for her Ph. D. work from Ministry of Tribal Afffiars GOI. Her Ph. D. thesis has been awarded with Best Ph.D. Thesis Award at International Conference organised by Navsari Agricultural University, Navsari. She is currently working as Senior Research Fellow in the Government of Gujarat Project at International Agribusiness Management Institute, Anand Agricultural University, Anand from last two years. She has published 23 research papers in NAAS, WOS, and UGC care listed journals and also published 03 book chapters. She has participated in 05 different national and international seminar & conferences.

Project Management & Agribusiness Entrepreneurship Development

Dilip Rasiklal Vahoniya
Assistant Professor & Head
Department of Agri-Entrepreneurship & Project Management
International Agri-business Management Institute (IABMI)
Anand Agricultural University
Anand, Gujarat

Prakashkumar Hasmukhbhai Patel
Assistant Professor
Department of Commerce and Business Management
Faculty of Commerce
The Maharaja Sayajirao University of Baroda
Vadodara, Gujarat

Alvirakumari Ashokbhai Rajwadi
Senior Research Fellow
International Agribusiness Management Institute (IABMI)
Anand Agricultural University
Anand, Gujarat

NIPA® GENX ELECTRONIC RESOURCES & SOLUTIONS P. LTD.
New Delhi-110 034

NIPA® GENX ELECTRONIC
RESOURCES & SOLUTIONS P. LTD.
101,103, Vikas Surya Plaza, CU Block
L.S.C. Market, Pitam Pura, New Delhi-110 034
Ph : +91-11-43860225, Mob.: +91 9717133558, 9540816132
E-mail: newindiapublishingagency@gmail.com
Website: www.nipaersources.com

Print ISBN: 978-93-58879-95-7
ebook ISBN: 978-93-58878-40-0

Composed and Designed by NIPA®.

Dedicated to
my inspiration & respectful late parents
father (Er. Rasiklal Chhaganbhai Vahoniya) and
mother (Jelamben Rasiklal Vahoniya)

(Dr. Dilip Rasiklal Vahoniya)

Preface

Agribusiness is a vital sector that drives economic growth, ensures food security, and provides livelihoods to millions worldwide. However, the success of agribusiness ventures depends on more than just agricultural knowledge—it requires strategic planning, resource management, and entrepreneurial skills. This book, ***Project Management & Agribusiness Entrepreneurship Development***, is designed to bridge the gap between theoretical concepts and practical applications, equipping aspiring entrepreneurs, agribusiness professionals, and students with the tools needed to navigate the complexities of the industry.

In today's dynamic business environment, effective project management is crucial for ensuring the sustainability and profitability of agribusiness enterprises. From conceptualization and planning to execution and evaluation, structured project management methodologies can help entrepreneurs maximize efficiency, mitigate risks, and achieve long-term success. This book integrates key principles of project management with agribusiness entrepreneurship, offering insights into business development, financial planning, market analysis, risk management, and innovation in the agricultural sector.

The chapters of this book are carefully structured to provide a step-by-step guide, beginning with fundamental concepts and progressing toward advanced strategies. Readers will explore case studies, best practices, and real-world applications that illustrate how successful agribusiness ventures are developed and managed. Whether you are a student, researcher, policymaker, or practitioner, this book aims to serve as a comprehensive resource to enhance your understanding of agribusiness entrepreneurship and project management.

We hope this book empowers you with the knowledge and confidence to create, manage, and sustain agribusiness enterprises that contribute to economic development and food security.

Dilip Rasiklal Vahoniya
Prakashkumar Hasmukhbhai Patel
Alvirakumari Ashokbhai Rajwadi

Contents

1

Introduction to Project Management

The term *project originates from the Latin word projectum derived from the Latin verb proicere*, "before an action" which originates from pro-, signifying precedence, something that occurs prior to something else in timing, and icere, "to do". The term "project" therefore initially signified "before an action. ". A project is a distinct kind of production system that produces a sole unit of product. Like, building a ship or a house, moving a department or a plant, developing a new product or performing a study are examples of such one-of-a-kind products.

In modern business and science, a project is a joint effort, encompassing research or design, that is carefully organized to accomplish a specific goal. The project can also be described as a collection of interconnected tasks to be performed within a defined timeframe and under specific budgetary and other constraints. Projects can be additional defined as temporary instead of enduring social systems or work systems that are formed by teams within or between organizations to achieve specific tasks under time constraints. And all must be expertly managed to deliver the on-time, on-budget results, learning, and integration that organizations need. A project in business and science is a provisional endeavor engaged in to produce a distinct item, service, or result. Basically, it is planned to attain a specific goal. The aim of a project is to achieve its goal and subsequently conclude. Some of the justifications or motivations for initiating a project can be:

1. A customer request or market demand
2. An organizational need
3. A technological advance
4. A legal requirement

Demand is the quantity of a good or service that consumers and businesses are willing and able to buy at a given price in a given time period. Market demand is the sum of the individual demand for a product from buyers in the market. A customer request, is a requirement from the customer at given price, quantity, and quality to a business that fulfills the requirements of the clients. An Organizational Needs Analysis is an efficient method to recognize any disparity between the abilities your business requirements and those your team possesses. It entails collecting data to pinpoint areas in which your employees can enhance their performance to their benefit and that of your business.

A technological advance, a modification to a product or service's production or delivery that lowers the amount of resources needed for production or delivery. For instance, a technological advancement in the wireless broadband sector may enable more effective use of limited spectrum, increasing wireless broadband's speed and capacity. Over time, there have been numerous alterations in the agricultural sector. While certain ancient techniques and technologies still exist, more effective, innovative, and efficient agricultural technological breakthroughs are opening up new possibilities and changing the way farming is done. This has altered the way crops are raised and produced more effective ways to manage resources. Technology has an unquestionable impact on agriculture today. Engineers and scientists are constantly putting in a lot of effort to create new technologies that address issues with farming, crops, and animal management.

The following list of **five technological developments** that have a significant impact on agriculture:

A. Precision Agriculture

Precision farming entails gathering data on crops and soil using GPS and other technical tools in order to optimize inputs (water, fertilizer, etc.) based on particular circumstances. Crop growth can be enhanced while simultaneously lowering wastage by monitoring these variables and adàpting to changes in them, such as moisture levels. It aids farmers in using inputs more precisely, cutting down on waste and saving money. One of the most popular agricultural technical developments, particularly in large-scale farming where every input counts. Precision farmers experience higher yields, better soil health, and a positive environmental impact. For instance, farmers may prevent over fertilizing the field, which can be wasteful and lead to disease, by using the technology that is already available to check soil health.

B. Industrial Automation

To accomplish operations like precise field seeding, planting, fertilizing, pesticide/herbicide spraying, and crop harvesting, robotics and other automated procedures are used. By improving productivity on farmlands, this technical innovation in agriculture has helped farmers to improve crop harvests. Drones can now be used to map crops, track agricultural growth, and enhance irrigation techniques. In order to acquire a bird's eye perspective of the landscape, evaluate fallow fields, or keep track on irrigation levels over a vast area, drones are also employed for aerial surveys. Drones are being used by more farmers to map their properties for ideal growing seasons, crop rotation plans, and harvest requirements. Robotics has facilitated the creation of machines that can milk cows, shear sheep, and other tasks in livestock production.

C. Automated Irrigation Systems

The automation of irrigation systems is a great example of how technology is affecting agriculture. These technologies have completely changed the way water is delivered to crops, increasing water distribution efficiency and agricultural production quality and quantity. Modern irrigation techniques efficiently deliver water when it is most needed. Better crop yields and more effective water distribution are made possible by this accuracy. Farmers in areas with water limitations brought on by drought or climate change stand to gain the most from this agricultural technology development.

The future is bright for farmers and their crops as irrigation becomes one of the main factors influencing agricultural success. Farmers who adopt this strategy may have an advantage.

D. Remote Monitoring of Crops Using Sensors

It is increasingly common to remotely monitor crops using sensors like drones and satellites. This enables farmers to keep an eye on their crops from the comfort of their own homes, increasing output by spotting issues earlier and enabling more effective use of water and fertilizer. Using an app or web browser, crop sensors allow farmers to remotely monitor their crops from any location in the world. With such agricultural technical innovation, producers may reduce labor expenses and boost crop yields, putting an end to the food shortage. Smallholder farmers can remotely monitor their crops using sensors in addition to large-scale farmers. According to a recent study, remote sensing could increase smallholder farmers' yield projections in Africa by up to 30% in accuracy. This will help these farmers make better decisions about their farming practices.

E. Genetically Modified Crops

One of the most significant technological developments in the agriculture industry is the utilization of genetically engineered plants. These plants have been modified to include particular qualities that will be advantageous to both farmers and consumers. For farmers raising specialty crops like fruits and flowers, they provide a wealth of advantages. These include improved nutritional value, endurance under unfavorable climatic conditions, tolerance to herbicides, and higher resistance to pests and illnesses. GMOs have dramatically decreased the amount of pesticides farmers must spray on their fields over the previous 20 years by up to 8.2% while improving crop production by 22%. The science is crystal clear that these agricultural technological advancements are a safe and useful tool for farmers, even though they may not always be popular with consumers. Planting GMO crops also helps preserve soil, reduce carbon emissions, and conserve water.

1.1 Project Characteristics

- It is temporary – temporary means that every project has a fixed beginning and a fixed end. The project always has a final time frame.
- A project creates single units, which are products, services, or results.
- A project creates a skill to perform a service.
- Project is always established in steps and continues in increments – Progressive Elaboration.
- Specific timeframe /One-time process/complex
- Limited by Budget/resources / Scheduling
- Exceptional specifications.
- Working across organizational boundaries
- Develop to resolve a clear goal or set of goals.

- Projects are customer focused
- Projects are responsible for the newest Product, services, and processes.
- The traditional management functions are applied to the Project.

What is Project?

A project is a unique attempt to produce a set of deliverables within clearly specified time, cost and quality constraints. A project is temporary in that it has a clear beginning and end in time, and therefore defined scope and resources. And a project is unique in that it is not a redundant task, but a specific set of operations planned to accomplish a singular goal. So, a project team often includes people who don't usually work together – sometimes from different organizations and across multiple geographies.

Projects include everything from the creation of a building or bridge to the relief efforts following a natural disaster. They also include the extension of sales into a new geographic market. Projects and operations differ primarily in that operations are ongoing and repetitive, while projects are temporary and unique. Generally, a project is a means of organizing some activities that cannot be addressed within the normal operational limits.

Projects are different from ordinary business activities as follows:

- The nature of projects is distinct. They don't involve a process that is repeated. While operational activities frequently include repeating tasks, each project that is done is unique from the last.
- Projects have an established timeline. The time frame within which deliverables must be generated to satisfy a certain client need is clearly stated for each project's start and finish dates.
- A project's budget has been approved. Projects are given a budgetary limit within which a deliverable must be created in order to satisfy specific customer requirements.
- Resources for projects are constrained. A predetermined amount of labor, tools, and materials are allotted to the project at the beginning.
- Incorporate a risk component. Projects incur business risk because they involve some level of uncertainty or an uncontrollable variable throughout the project.
- Bring about a good transformation. Typically, a project's goal is to enhance a company through the execution of business change.

Difference Between Project and Programs

Definition of a Program

A program is a portfolio comprised of many projects that are managed and coordinated as one component with the objective of achieving (often intangible) results and benefits for the organization.

Definition of a Project

A project is a temporary organization created to carry out predetermined time, cost, and quality limits while producing particular (typically physical) outcomes. An

(Executive) approved business case that balances the project's costs, benefits, and risks should always be used to define, carry out, and evaluate a project. Change control should be used to manage the business case for the project. The following Table-1 and Table -2 indicates the difference between the Program and project and detailed difference between Project and Programme respectively.

Table 1.1: Difference between Program and Project

Program					
Project		**Project**		**Project**	
Activity	Activity	Activity	Activity	Activity	Activity
Activity	Activity	Activity	Activity	Activity	Activity
Activity	Activity	Activity	Activity	Activity	Activity

Table 1.2: Difference between Project and Programs

	Project	**Program**
Objectives	Outputs – physical; relatively easy to describe, define and quantity; tending towards objective.	Outcomes – often intangible; difficult to measure; benefits often based on changes to organizational culture and behaviors; introducing new competences into the organization; tending towards subjective.
Scope	Strictly restricted; tightly defined; not likely to be subject to material modification during the life of the project.	Not tightly defined or restricted; likely to modification during the life cycle of the program.
Duration	Relatively short term; naturally, three to six months.	Relatively long term naturally eighteen months to three years.
Risk profile	Project risk is relatively easy to detect and manage. The project failure would result in a fairly limited impact on the organization relative to program risk.	Program risk is more complex and potentially the impact on the organization if a risk appears will be greater relative to project risk. Programme failure could result in material financial, reputational or operational loss.
Nature of the problem	Clearly defined.	Often disagreement between key stakeholders on the nature and meaning of the problem.
Nature of the solution	A comparatively limited number of potential solutions.	A considerable amount of possible answers typically exists with frequently differing opinions among stakeholders regarding the best solution.
Stakeholders	A relatively small group of stakeholders.	A significant number of various stakeholders; likely conflict between them regarding the interpretation of the issue & the preferred solution.

Relationship to environment	Environment inside which the project takes place is understood and relatively stable.	Environment is dynamic; and programme purposes must be handled within the framework of the moving environment within which the organization operates.
Resources	Resources to deliver the project can be reasonably projected in advance.	Resources are controlled and limited; there is competition for resources between projects.

Types of Project

There are generally five different types of projects, such as those that produce artifacts through construction, knowledge through research, change through reengineering, business relationships through procurement, and working processes through business implementation. Each of these projects provides a distinct answer to the subsequent questions:

- Where do we start?
- How do we know when to stop?
- When can (should) we evaluate the results?

Let's discuss all the five projects in detail one by one, Table 3 indicates the types of project with justification.

Construction Projects

The project produces an artefact. The value generated by the project is embedded in the artefact. The artefact may be a complex system with human and mechanical components.

Examples

- Warship
- Jubilee line extension
- Millennium dome
- Customer call center
- Method guidebook
- IT system

Research Projects

The project produces knowledge. Models, patterns, or patents are some formal representations of the knowledge. Or the information might be included in a procedure or artifact.

Examples

- Enterprise modeling
- Creating a model of the British economy
- Creating a new variety of wheat
- Creating innovative project management techniques

- Military intelligence/cracking codes
- The project's analysis, testing, quality assurance, or evaluation phases

Reengineering Projects

The project produces a desired modification in some system or process.

Examples

- Converting pounds to euros
- Implementing PRINCE (Projects IN Controlled Environments) project management techniques in a large firm. Renumbering the UK telephone system.
- Creating and setting up an intranet.
- Irrigation initiatives

Procurement Projects

The project results in a contractually-based business relationship with a chosen supplier for a specified good or service that is based on a set of specifications and/or a specified specification procedure.

Examples

- Contracting out a particular building or research endeavor.
- Outsourcing an entire corporate operation (like IT).
- Adding extra regulations and restrictions to an already-regulated sector.

Business Execution Projects

The project produces an operationally impressive process. The value generated by the project is embedded in the process.

Examples

- Developing a new business process to repackage and exploit existing assets.
- Installing e-commerce

Table 1.3: Types of Projects

Types of Projects	**Start**	**Stop**	**Evaluate**
Construction Project	• Start with a set of requirements. • Start with a defined solution.	• Stop when the artefact is "complete". • Stop when the requirements are satisfied.	• Evaluate on delivery of the artefact. • Evaluate over the lifetime of the artefact.
Research	• Start with a hypothesis. • Start with a problem.	• Stop when the time runs out. • Stop when we detect diminishing returns.	• Evaluate when the knowledge is confirmed or dis confirmed by later work. • Evaluate when the knowledge is used by later work.
Reengineering	• Start with a problem. • Start with an opportunity. • Start with an (imported solution).	• Stop when we seem to be ahead of the game. • Stop when some higher process changes the game we're playing.	• Evaluate at any time.
Procurement	• Start with a set of requirements. • Start with a defined solution.	• We construct a tender document that is "complete". • We sign a contract with the supplier who seems to be ahead at the end of the tender period.	• Over the lifetime of the contract. • On completion of the contract.
Business Implementation	• Start with an opportunity. • Start with a business concept.	• Stop when the process is operational. • Stop when the process has been running smoothly for a defined period. • Stop when the business benefits are starting to become visible.	• Evaluate when the process has been running smoothly for a defined period. • Evaluate when the business benefits are starting to become visible. • Over the lifetime of the process.

Stages of A Project and How to Complete Them

Stage 1: Determining the goals of the project

Garbage in equals garbage out. If manager do not start a project with the correct goal(s), it is not likely for manager to accomplish the goal(s) in time. This activity must be a team effort to allow everyone the chance to contribute and "buy-in" to what is about to happen. Each team member should be aware of what has to be done by the end of this stage.

The project should conclude with a paper that specifies the aims and includes a brief statement that provides some explanation. A definition of success should be incorporated in the assertion. This assertion describes the crucial few conditions that must be met in order for the aim to be deemed successful. The phrase "vital few" refers to the bare minimum needed to perform well. To ensure that there is no doubt regarding whether the aim has been achieved or not, success must be defined. Additionally, it stops teams from making pointless improvements to a task that has already been satisfactorily completed.

Stage 2: Determine project tasks/activities

The tasks necessary to complete each objective or subset of goals should be matched to those goals. The simplest way to achieve this is to write the tasks next to the objectives on the right side of a piece of paper. The group should come to an understanding that the specified task will achieve the objectives as outlined in the definitions for success the team established in the earlier stage.

Stage 3: Determine and verify resource requirements

Resources are items that are necessary to finish the project's objectives. Resources could consist of, but not be limited to:

- People
- Money
- Space
- Time
- Computers
- Software
- Others

Stage 4: Identify risks and create mitigation (backup) plans:

There is always a certain element of risk (uncertainty) associated with projects, which could cause issues and unpleasant shocks. Surprises require more time, effort, and resources than originally anticipated. Risk management can aid in lowering risk's likelihood and effects. Risk management is crucial since it allows the team to finish the project with the least amount of difficulty.

You will be expected to contribute as a qualified scientist, engineer, or project manager to finish the assignment. In general, excuses—real or made up—don't win anyone over. For instance, it is all too simple to shift blame when a surprise issue slows a project by using language like "I did my part, but the other guy didn't" or "We had an unexpected problem that caused us to fall behind." Managers, educators, and CEOs often have negative opinions of those who use victimization language. Therefore, even if they are correct, people who rely on these kinds of statements frequently do not succeed in this world. You must learn how to handle unexpected and hardship instead of allowing them to dominate you. Each person who joins a team commits himself to the success of the TEAM. It is your duty to assist a team member in any manner you can if you are anticipating their arrival to finish a task and they are running late. If you cannot help them, make sure they receive the assistance they need to complete their assignment. The "victim" word makes you appear uncooperative rather than absolving you of any culpability.

Although risk management can be a complicated process, we will employ a straightforward approach for our objectives. List the practical dangers linked to each of your responsibilities in a column on the right side of the tasks themselves. Create a backup strategy for each risk that, when used, will keep you on track. Risks that are so significant they could "kill" the entire enterprise merit special consideration. Throughout the project, One of the members of the group ought to take responsibility for monitoring this risk.

Stage 5: Develop a schedule

A Gantt chart is a schedule that displays the tasks, the individuals in charge of them, and a timeline. Gantt charts are beneficial as they allow the team readily identify roles and view the project's architecture (structure). People can use the gantt chart as a planning tool to visualize how tasks are interdependent by using it as a document. A gantt chart's fundamental structure is a list of tasks on the left, followed by the start date, the number of days needed to finish it, and the finish date. There should be one or more owners assigned to each task. A graphic depiction of the task length in connection to the project time line is positioned to the right of the text listing.

Why Projects Win?

Though project has different requirement and it is a onetime operational activity but still it required different assistance from different point to be successes. The point is as follows:

- Executive level project sponsorship
- A sound project charter
- The correct combination of team members
- A sound decision-making structure
- Good communication
- Team members are collaborating to achieve common objectives

Let discuss each and every point in detail

Project Sponsorship at executive level

Executive sponsor (sometimes called **project sponsor** or senior responsible owner) is a role in project management, normally the senior member of the project board and often the chair. A senior executive in a company, frequently at or just below the board level, who is accountable to the company for the project's success will serve as the sponsor. For any project to be successful, the executive level must provide a strong positive response and support. The executive sponsor must create a common ground on which all stakeholders may agree and work together to deliver while considering the varied and frequently conflicting stakeholder interests. The executive sponsor's responsibility is to negotiate the option with the strongest business case in order to provide the investor with the highest value feasible. The sponsor's responsibility begins from the very beginning of the project. Sponsors play a key role in the creation of the initial scope and charter in addition to guiding the project through the selection process until it is finally authorized.

A sound project charter

It specifies the project goals, identifies the key stakeholders, and defines the project manager's authority. It also gives a basic definition of duties and responsibilities. For the project's future, it acts as a source of authority. The project charter typically includes the terms of reference.

The ideal combination of team members

In the majority of firms today, employees work in teams or groups. This means that managers need to be adept at managing both individuals and groups. As a manager, It is your duty to build an efficient team in which each member adds value, does their part, and values and supports one another. You must continuously promote teamwork and collaboration among the individuals you supervise if you desire to achieve success in this. The project's success relies on every individual in the team. having the abilities to do their jobs. The entire team suffers when there are skills gaps. The team succeeds when the members have the correct combination of skills.

A sound decision-making structure

Organizational structures in business were developed to make decision-making simpler. Because executive officers determine the company's goals and direction, the standard organization chart shows a hierarchy of power with the chairman, CEO, and other executive officers at the top. Applying these objectives to their functional areas or departments, which are ordered in the sequence of proper information flow from management, is the responsibility of supporting layers of managers beneath the executive level. An organizational structure that separates a company's operations into specialized departments gives the managers of those departments the authority to solve issues and develop production and process efficiencies that are tailored to their specific departments. Additionally, it establishes a corporate management level where enterprise knowledge is created, evaluated, and kept for the department's continued effective operation as well as for simple access by those in higher management who are accountable for business planning.

Good communication

The subsequent skills of the project manager are shown in Figure 1 to deal with both high-level management and low-level employees, increasing the likelihood of the project's success.

Figure 1.1: different communication skills

Exchanging information is only one aspect of communication. It's important to comprehend the motivations and feelings that underlie the information. Effective communication requires reciprocity. It's important to listen carefully in order to understand what is being said fully and to make the other person feel heard and understood. This will ensure that your message is received and understood just as you intended. Effective communication involves more than just using words; it also involves nonverbal communication, attentive listening, stress management in the moment, the capacity to communicate assertively, and the ability to recognize and understand both your own emotions and those of the individual you are conversing with. You may strengthen relationships with people and enhance teamwork, decision-making, and problem-solving by communicating effectively. It helps you to convey even disagreeable or challenging messages without arousing hostility or undermining confidence.

Team members are collaborating to shared goals

Teams are kept together via training and team building exercises used by businesses and organizations of all sizes. Goals are an organization's broad, overarching objectives in business. By focusing personnel on shared objectives, a company can enhance working conditions and the bottom line. One of a team's distinguishing traits is its ability to work together to achieve shared objectives. It is impossible to be productive and compete with firms that are more goal-focused when individuals are working for their own goals.This means that effective team building requires open communication and consistent objectives. At the time of hiring, business executives must make sure that every employee is aware of the organizational culture, overarching goals, and individual personal aspirations. To keep shared objectives at the forefront of employees' thoughts, managers must keep an eye on both team development and individual growth.

The qualities of successful teams should be promoted via team building exercises that aim to bring personnel together around a common objective. These include participation from all team members, self-motivation within the team, and shared accountability for errors. Strong communication and a willingness to clarify things for one another are other characteristics of effective teams. Business executives should evaluate each of these components in order to invest time and resources in fostering the skills that a goal-directed team lack.

Why Projects Fail?

Project also fail because of some reason which we already discussed under the above topics why project win? We mention some more reasons for project failure are as follows:

- Inability to connect the project with the goals of the organization
- Unrealistic assumptions
- Insufficient executive backing
- Insufficient project management
- Failure to progress past personal and character conflicts
- Politics within the team and within the organization

Why Projects are Important?

- Shortened product life cycle (PLC).
- Limited product launch window
- Increasingly complex and technical products
- Emergence of global market
- An economic period market by low inflation

Let discuss each point one by one in detail

Shortened product life cycle

The day when a business could launch a new product and count on a year of market dominance is over. More often than not, the lifespan of new items is expressed in terms of months or even weeks, as opposed to years. To detect trends in electronics, computer hardware, and software, one simply needs to look at a single device. We are increasingly noticing similar signs in traditional service sector companies as well, who have also realized the importance of agility in providing and upgrading new services at an accelerated rate. Companies are being forced to create new products at a faster rate than ever before due to heightened competition and shifting consumer preferences. Customers appear to desire more features and quicker processors more and more, which seems to be the case in the technology industry in particular. Businesses are going through some substantial changes as a result of shorter product life cycles. Companies are forced to simplify their networks by the requirement to replace products every few years in order to meet demand.

Limited product launch window

The opportunity's nature is the other time-related concern. Organizations must adopt a proactive approach to the timing of product introductions since they are aware that there is no optimal time to launch a new product. Smart businesses, for instance, are already planning the optimal time to introduce product 2, either as an upgrade to existing products or as a whole new offering, while simultaneously securing the benefit from the successful sale of product 1. The best launch possibilities are evaluated in terms of months due to the fierce (aggressive) competition. Miss your launch window by even a week and you are operating dangerous automobiles.

Increasingly complex and technical products

The modern world is somewhat complicated, and there is fierce market competition. The public wants increasingly technologically advanced items, which makes it harder for businesses to create new products while still satisfying their target market. items are also becoming more complex and more technology is used to develop them. We constantly strive to make newer models of our consumer items better, bigger, faster, and more complicated than their predecessors. To meet this need, businesses in this line consistently labor. Additionally, many businesses concentrate their research and development efforts on future demand and create goods based on that study in an effort to meet that demand.

Emergence of global market

We have witnessed the creation of markets for practically all types of goods and services in the past. A business now faces new obstacles as a consequence of the increased economic globalization and improved ways to communicate with suppliers and consumers fast. Challenges can also be opportunities for businesses that can adapt to this new environment swiftly. Project management approaches give businesses the opportunity to connect several business partners, react swiftly to market demands and supplier needs, and remain responsive to the rapid change in customer taste. Future successful organizations will identify and learn to take advantage of the opportunities presented by a global business environment through the use of project management.

An economic period market by low inflation

The fact that inflation has been kept in check is one of the main indicators of economic health. Particularly in the US, low inflation must be the primary cause of the stock market boom of the past ten years as well as continuous economic growth. Unfortunately, low inflation also affects the company's ability to pass on cost increases to customers and sustain profitability. A company cannot maintain its profit margin by simply raising the cost of its goods or services. Future successful businesses will be those who increase earnings through internal process optimization and who save money by outperforming their competition.

Characteristics of Project Manager

- Interpersonal skills
- Organizational skills
- Communication skills
- Problem-solving skills

Interpersonal skills

Interpersonal skills are life skills that we use every day to interact and interact with other people, both individually and in groups. People who have worked to develop strong communication skills tend to be more successful in their professional and personal lives.

Manager required communication skills to deal with people. Interpersonal skills are life skills that we use every day to interact and interact with other people, both individually and in groups. People who have worked to develop strong communication skills tend to be more successful in their professional and personal lives. Employers often look to hire employees with "strong interpersonal skills" – they want people who work well in teams and can communicate effectively with colleagues, customers, and clients. Manger required the following communication skills including:

- **Verbal communication** – what we say and how we say it.
- **Non-verbal communication** - body language is an example of what we communicate without words.
- **Listening skills** - how we interpret both verbal and non-verbal messages sent by others.
- **Negotiation** - working with others to find a mutually satisfactory outcome.
- **Problem Solving** - Collaborating with others to identify, define and solve problems.
- **Decision making** - examining and analyzing alternatives to make healthy decisions.
- **Assertiveness** - to communicate freely about our values, ideas, beliefs, opinions, needs and desires of.

Organizational skills

Alongside communication and computer abilities, organizational skills are one of the most significant transferable job skills an employee can have. Companies need employees who can stay organized and focus on the projects. However, company managers must also organize the work of their employees.

Communication skills

Good communication skills in the workplace means being able to convey information to people clearly and simply in a way that means things are understood and done. It means communicating and delivering your messages and being able to read your audience.

Problem-solving skills

The process of working through the details of a problem to find a solution. Problem solving may involve mathematical or systematic operations and may be a measure of an individual's critical thinking.

2

Project Management, Project Life Cycle and Project Feasibility

Project management involves the use of knowledge, skills, tools, and techniques to design activities to fulfill project requirements. In other words, project management encompasses the skills, tools and management processes needed to successfully implement a project.

Project Management comprises

- **Skill:** Expert knowledge, skills and experience are necessary to reduce the risk level of a project and thereby enhance the likelihood of its success.
- **A set of tools:** Project managers use a variety of tools to improve their chances of success: Examples include document templates, records, planning software, modeling software, inspection checklists and inspection forms.
- **Sequence of processes:** Various Management methods and procedures are necessary to oversee and regulate the time, cost, quality and scope of projects. For example, time management, quality management, cost management, change management, problem management and risk management.

Project Management is adequate when or when should you use project management techniques

- The objective is clear
- The objective is complex in scope
- Different organizations participate
- Quick responses to changes are required
- The tasks are highly interdependent
- The reward / penalty is high

Project Triple constraints

Understanding the Triple Limit of Project Management. All projects are carried out with specific limitations - traditionally these are cost, time and volume. These three factors (often referred to as the "triple limit") are represented as a triangle (see Figure 2). The Classic Triple Constraint From a business perspective, a project is a time-bound effort designed to move an organization from one level of measured performance to a higher level of measured performance. To determine whether we have achieved the project goal, we need good measurement methods. When we measure the wrong things, our focus and attention shift significantly away. The traditional triple bottom line for assessing the project's success is insufficient as it fails to evaluate how effectively we have achieved success in relation to commercial opportunities. It measures the wrong things and distracts us from the real possibilities. Let's study the classical Triple Constraint.

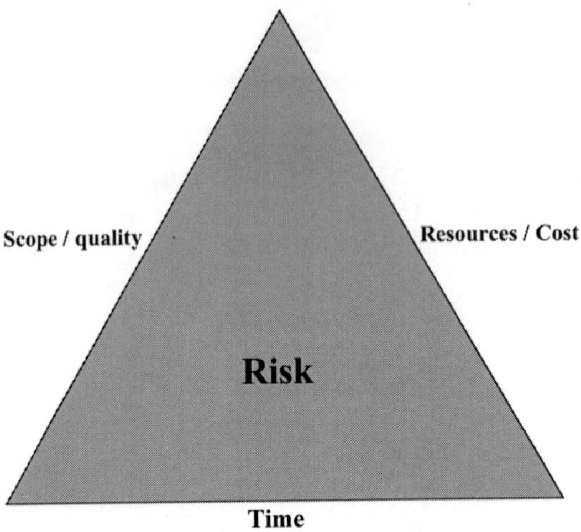

Figure 2.1: Classical Triple Constraint

Triple Constraint: A framework that evaluate competing demands. Project managers often talk of a "triple constraint" – project scope, time and cost – in managing competing project requirements (PMI, 2004, p377). Let's examine in some depth the basics of the triple constraint. A time and cost constrained process to achieve a series of specified deliverables (the scope to meet the project's goals) according to quality standards and specifications.

Cost: Unless you're lucky sufficient to be engaged in a project supported by an endlessly affluent backer, you intuitively grasp the expense and resource constraints. Money is limited. An infinite number of employees cannot participate in the project. "Money is no object" only implies that the spending limit can be the most adaptable, not that there are infinite dollars or euros to utilize. Here are several factors and examples of spending limits.

Budgeted expenditures: Expenditures in June are capped at $110,000.

Personnel constraints: 3 Scientists and 1 Technician, 14 staff-years of work effort

Equipment: 12 Spectrometers, 5 forklift vehicles

Consumables: 3,000 barrels of paint, 160 tons of concrete mixture, 500GB of storage capacity

Overhead/Administration: The software for Project Management needs to be Brand X, version Y.

Intangibles: Utilization of political influence, privileges, bargaining, compromises

Time: Even if an "extraordinary marvel, extraordinary marvel " happens and you unexpectedly acquire limitless resources for your project, the project's definition restricts you from utilizing an infinite amount of time. Remember that a project has clear beginnings and ends. So, your project has a deadline by which it absolutely must be finished. And although that deadline is established, differences in the timing and length of related tasks and subtasks will certainly alter that final date. Scheduling and sequencing methods like CPM and network diagrams, along with the straightforward, consistent use of a collaborative project calendar, assist project managers in handling deadlines. Below are some aspects of the time constraints:

Scope: In contrast to the time and resources constraints that the project "uses up" as you (hopefully) progress toward the project's outcome, scope is a boundary established by the vendor/supplier/vendor and constrained by the contract with the client/customer/buyer in contract form.

Quality: Below is the present definition of quality as provided by the American Society for Quality (ASQ):

Quality: A term that is subjective and for which each individual has their own interpretation. In technical contexts, quality can refer to two definitions: 1. Attributes of a product or service that influence its capability to meet expressed or presumed requirements. 2. Flawed product or service.

We have agreed to define project quality (so far) as the productive outcome of the project for the client that most effectively optimizes resource utilization and time and meets all comprehensive requirements. Quality is defined as "what the customer wants". Some experts believe that the "triple limit" should be the "fourth limit", with "customer acceptance" or quality as the fourth limit. Some have mentioned that there are five restrictions, with risk being the fifth. So, there are conflicts in the world of boundaries, but when it comes to deciding project management boundaries, the world of conflicts seems to have no boundaries.

Laws of Project Management

- No major project is ever installed on time, under budget, or with the same staff that started it. You are not the first.
- When things are going well, something goes wrong.
- If things seem to be getting better, you've forgotten something.
- No system has ever been fully tested. Attempts to debug the system inevitably lead to new errors that are even harder to find.
- A poorly organized project requires three times more time than anticipated
- A meticulously organized project requires only double the time.
- Project teams hate reporting progress because it clearly shows their progress.

What is Project Management?

A project is usually initiated by a perceived need, problem or opportunity within the organization. Being self-employed, it has a start and finish, budget, time and resource constraints and involves a purpose-built team. Project teams consist of many different team members, such as (product or service) end users/customers, information technology (IT) representatives, project managers, business analysts, trainers, project sponsors, and other stakeholders.

Project management is a discipline in which all the various resources and aspects of a project are managed so that the resources produce all the necessary results to complete the project within specified boundaries, timeframe, and budget constraints. These are established at the beginning of the project, and when the project begins, all stakeholders and team members possess a clear comprehension and agreement regarding the process, methodology and expected results. A good project manager uses a formal process that can be reviewed and used as a project plan, and this is accomplished through a project management approach.

The Project Management Context

Projects and project management occur within a larger environment than the project alone. The project management team needs to grasp this wider context to select the appropriate life cycle stages, processes and tools and technologies for the assignment. Context is the combination of existing environmental, political, socio-economic and business factors that particularly affect the way a project is executed or the result of a project.

Programme Management in Relationship to Project Management Context

Project management is, as we described before, a discipline that, in its traditional form, focuses on defining, planning and implementing specific objectives. Project management exists in a wider context that encompasses program management, portfolio management and project management. A program is a collection of related projects that are managed in a coordinated manner to provide benefits and control not available when managing individual projects. There is frequently a ranking of strategic plan, portfolio, program, project and sub-project, where a program consisting of several interrelated projects contributes to the accomplishment of the strategic plan.

The increasing use of projects to effectuate considerable organizational transformation has led to widespread interest in program management. By grouping, directing and initiating new projects, programs create a framework for projects, but they are also exposed to external influences and development. Emerging thinking on best practices in program management has focused on the internal context, namely the organization of projects and resources to achieve desired strategic and/or synergistic benefits. Relatively overlooked, although critical, is the need to modify, embed and adapt the program to the changing needs of the organization. Programs form the bridge between projects and organizational strategy, program management takes many of the characteristics and skills of strategic change management and organizational development. Program managers and leaders play an important role as mediators and manipulators of external contexts and pressures. They must accept and deal with the ambiguities, tensions and interactions between the need for specific actions and strategic openness, options and rapid response.

Programs may include elements of related work outside of the scope of the discrete projects in the program. For example:

- A new car model program can be divided into projects to design and update each major component (e.g., transmission, engine, interior, exterior) because continuous production takes place on an assembly line.
- Many electronics companies have program managers who are responsible for coordinating both individual product releases (projects) and multiple releases over time (ongoing operations). Programs also include some recurring or cyclical projects.
- Utilities such as Eskom or Transnet often talk about an annual 'construction programme', which is a series of projects that build on previous efforts but are designed to achieve a long-term goal or strategic objective.
- Many non-profit organizations have a fundraising program to obtain financial support for several separate projects, such as membership or auctions. It happens over time and is related to the tactical objectives of the organization.
- Publishing a newspaper or magazine is also a publication where every single issue is managed as a project. This is an example where general activities can become "project management".

Unlike project management, program management is the centralized, coordinated management of a project team to achieve the strategic goals and benefits of the program.

Project Management vs. Ongoing Management

Is project management an art or a science? If it's science, isn't it fair to say it's smart science. So, what's all the fuss about project management? If this is common sense, why can't we simply extend the management disciplines we already apply to the day-to-day management of businesses and corporations and apply them to the execution of our projects? The answer lies in understanding the difference between managing an ongoing business and managing a project.

Ongoing Management

The continuous general management of the organization is marked by recurring phases of planning, implementation and management of basic business processes. Cycle after cycle, incremental improvements are made to improve efficiency, reduce costs, improve product quality, etc. General (continuous) management is a broad subject that deals with all aspects of management, including

- Finance and accounting, sales and marketing, research and development, manufacturing and distribution
- Strategic planning, tactical planning and operations planning
- Organizational structures, organizational behavior, staff development, staff management, payroll etc.
- Management of working relationships through motivation, delegation, supervision, conflict management, team building.

The Japanese philosophy of Kaizan best describes this type of management with its slogan of "*there is no best – only better, and str*ives for continuous improvement in the repetitive processes.

Examples of ongoing general management are,

- Managing a supermarket, hospital or hotel
- Manufacturing shoes, cars or wine
- Managing an international airport or convention center
- Managing an engineering firm or county government• managing a company. airline or taxi fleet or restaurant

Every aspect and process of the business that supports the company is constantly reviewed and improved with the ultimate goal of improving profitability and increasing shareholder returns.

Phases of Project Life Cycle

The Project Lifecycle consists of four phases:

1. Initiating
2. Planning
3. Executing
4. Closing

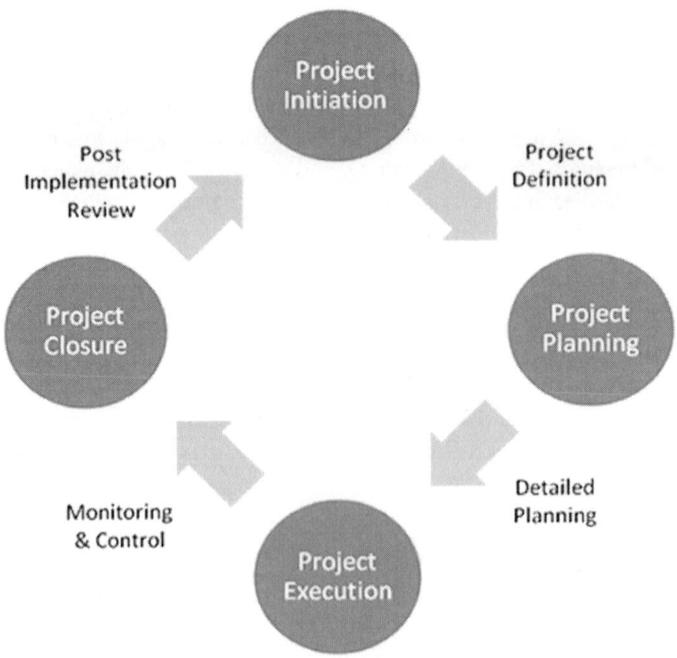

Figure 2.2: Phases of Project Life Cycle

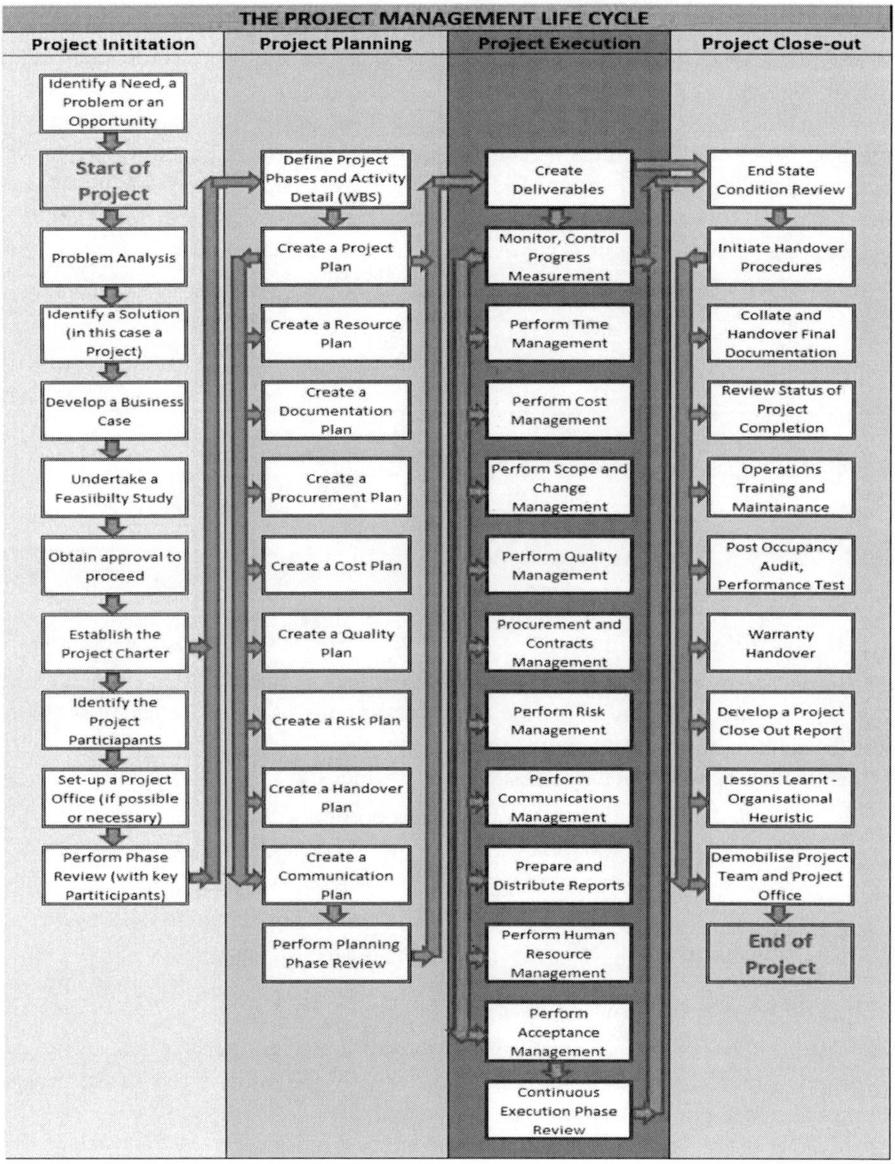

Figure 2.3: Phases of Project Management Life Cycle

Let's have a detail explanation about these above four steps one by one

1. Project Initiation

The launch phase is the initial phase of the project. In this step, a business problem (or opportunity) is identified and a business case is defined that offers different solution options. Next, a feasibility study is conducted, in which the probability of each solution option to solve the business problem is determined, and the final recommended solution is presented. Once the recommended solution is approved, a project is initiated to deliver the approved solution. A "mandate" is drawn up, outlining the goals, scope and structure of the new project, and a project manager is designated. The project manager begins to recruit the project team and establishes the project office. Approval then moves to the detailed design. The initiation phase basically involves "launching" the initiative. This represents the stage where a business problem or opportunity is identified, a solution is agreed upon, a project is established to develop the solution, and a project team is appointed. The following diagram describes the steps taken:

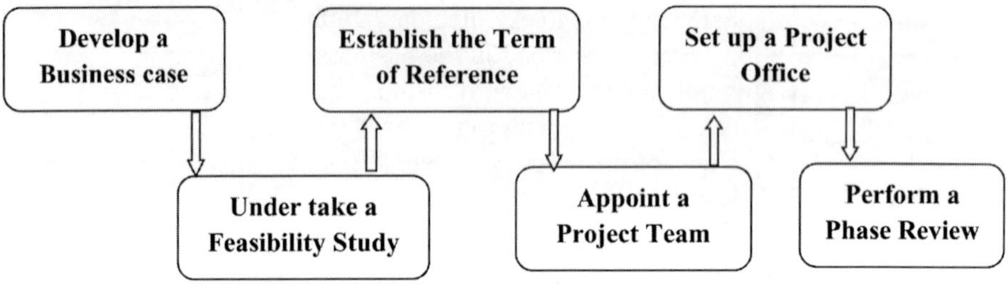

Figure 2.4: Phases of Project Initiation

1.1. Develop Business Case

The business case validates the initiation of the project. It contains a description of the business problem or opportunity, the costs and benefits of each alternative solution, and the solution recommended for approval. Once a business problem or opportunity is identified, a business model is created. This includes:

- A detailed definition of the problem or opportunity
- An analysis of the possible solution options available. The potential benefits, costs, risks and problems of each option are documented. An official feasibility study can be ordered if the feasibility of a certain solution option is not clear.
- Recommended solution and general implementation plan.

When do we use a Business Case?

The business case template is utilized whenever there is a need to justify project expenses. Completing the business model is usually the initial phase in the project life cycle. Once the business case model is ready, it is submitted to the sponsor for approval. A business case is often referenced during a project to assess whether it is presently on course. And upon the conclusion of the project, success is measured by the ability to attain the objectives outlined in the Business Case. So completing the business justification is vital to the achievement of the project.

1.2 Perform A Feasibility Study

What is a Feasibility Study?

A project feasibility study is an assessment in which all possible solutions to a specific business problem or opportunity are documented. Feasibility studies can be performed by any company, project or team and are a significant component of the project life cycle.

When to use a Feasibility Study?

The aim of the feasibility study is to determine the likelihood that one or more solutions will fulfill the specified business needs. In other words, if you are uncertain whether your solution will deliver the desired result, a study on project feasibility can assist you gain that clarity. During the feasibility study, various "estimation methods" are applied. The outcome of the feasibility study is a validated solution for execution.

A formal feasibility analysis can be requested at any stage of (or after) the creation of the business model. The objective is to examine the likelihood that a certain solution variant will achieve the benefits described in the business case. The feasibility study also investigates if the anticipated costs are sensible, the solution is attainable, and the risks are manageable and/or potential problems are avoidable. Let us examine the comprehensive explanations of the feasibility study as outlined below.

Feasibility actually refers to whether or not the concept will be successful. He knows in advance if the suggested product/service possesses a substantial market, what the investment needs would be and where to get financing, if and where it is possible to find the essential technical expertise to transform the concept into a tangible item and so forth. In other words, a feasibility study encompasses an ex-ante (prior to the establishment of the company) examination of the company's operations, financial, personnel and marketing aspects. So, you can simultaneously read this lesson along with the lessons related to marketing, finance, etc. to gain a clearer insight into the concerns at hand. Below we provide a concise summary of the challenges impacting different facets of the viability of the suggested project. At this point, you should understand that feasibility is a multidimensional notion; In other words, the project has to be doable not just from a technical standpoint, but also economically and commercially. In addition, there is always the possibility that a technically feasible project may not be financially viable. For example, you can establish a dust-free factory in Rajasthan; however, it is more economically sensible to do so in Chandigarh/Bangalore. So, as we look at different aspects of practicality one by one, don't be deceived into thinking that there is a sequence and no interdependence.

Objective of the feasibility

After completing this lesson, the reader ought to be capable of

- Conduct a feasibility analysis of the suggested business concepts concerning
  - Marketability
  - Technical viability
  - Funding
  - Legalities
- Prepare a business plan
- Understand basic startup problems

There are various kinds of Feasibility which are as follows

A. Market Analysis
B. Financial Analysis
C. Technical Analysis
D. Economic Analysis
E. Ecological Analysis
F. Legal and Administrative Analysis

Let's take a detailed look at each feasibility individually.

A. Market Analysis

A market, regardless of whether it is a physical location, serves as a space for engagement between buyers and sellers. From the seller's point of view, the market analysis is based mainly on the total demand for the planned product/service in the future and the anticipated market share. The success of the planned project clearly relies on the ongoing assistance of the customers.

However, identifying the product/service market is quite challenging. After all, the whole universe cannot be your market. You need to carefully segment the market according to certain criteria like geographical scope, demographic and psychological profile of potential customers, etc. This is a study to find out who all your customers are. For that you need information:

- Trends in consumption.
- Historical and current supply situation
- Opportunities and limitations in production
- Competition in imports and exports
- Cost structure
- Responsiveness of demand
- Consumer actions, goals, drives, perceptions, choices, and needs
- Channels of distribution and marketing strategies being implemented
- Regulatory, operational, and legal limitations affecting the marketing of the Product

B. Financial Analysis

The goal of the financial analysis is to determine whether the suggested project is economically feasible in the sense that it can cover the debt service burden and if the suggested initiative satisfies the anticipated returns of capital producers. When making a financial assessment, certain aspects must be considered, for example:

- Expenditure on investment and project expenses
- Funding methods
- Projected profitability

- Break-even analysis
- Project cash inflows
- Investment value assessed based on different merit criteria
- Estimated financial status

C. Technical Analysis

The matters related to evaluating the technical analysis of the suggested project can be grouped into those concerning inputs, throughputs, and outputs.

Input Analysis

Input analysis mainly concerns the recognition, measurement and assessment of project inputs i. e. , equipment and resources. You must ensure that inputs of the right type and quality are accessible at the appropriate time and at the suitable price throughout the duration of the project. You must enter into long-term contracts with potential suppliers; in numerous instances you must endure your supply sources. When Macdonald came to India, they developed sustainable sources of potatoes, lettuce and other ingredients for their burgers. The activity related to developing and maintaining sources of supply is called supply chain management.

Throughput Analysis

It refers to the production/operations you perform on inputs to add value. Generally, received batches go through the conversion process in several production steps. Where the facilities are located, what the order would be, what the layout would be, what the quality control measures would be, etc., are things you will learn more about in the coming hours.

Output Analysis

This applies to the physical characteristics of the product - color, weight, length, width, height; functional properties; properties of chemical materials; and compliance standards like BIS, ISI and ISO etc.

D. Economic Analysis

Economics is the examination of profitability. Regarding the practicality of the investigation, the business owner is worried about whether the funds expenses and the cost of the item are justified compared to the cost at which it is sold in the market. For instance, in theory, silver might be obtained from silver bromide (a substance utilized in the processing of X-ray and photographic films); nevertheless, the expenses associated with mining are so elevated that it would not be financially viable feasible. Similarly, the operating costs of solar electricity were unreasonably high until recently. This break-even analysis goes into the financial calculations of the break-even analysis we covered in relation to the financial analysis. Here, It is also beneficial to differentiate between economic and commercial feasibility. While economic viability influences the unit cost of the product, business feasibility indicates whether sufficient units will be sold.

In addition to the above cost-benefit analysis, which we also refer to as a private cost-benefit analysis, it is also beneficial to perform a so-called societal cost-benefit Analysis (SCBA). For example, an entrepreneur can receive subsidized electricity, in which scenario the private expenses would be less than the societal expenses. Also, exporting units earn a valuable exchange rate, resulting in social benefits more than private income. Often, an SCBA-worthy project can gain more traction with supporting agencies.

E. Ecological Analysis

In recent times, environmental concerns have gained significant importance, especially in projects with a significant ecological impact, such as power plants and irrigation systems, but also in industries that pollute the environment (such as wholesale pharmaceuticals, chemicals and leather processing). The subsequent concerns are typically dealt with: - What is the probable environmental damage of the project? - What are the expenses associated with corrective actions to guarantee that the environmental harm stays within permissible bounds?

Table 2.1: Match the following

Sr. No.	Issues	Type of Analysis
1.	Availability of know-how with the business owner or to be obtained	Economic
2.	Capability to fulfill the obligation of managing debt	Technical
3.	The impact of the assignment concerning the standard of social costs-benefits	Ecological
4.	Identification, quantification, and assessment of project inputs	Financial

F. Legal and Administrative

Consider the plight of an entrepreneur who has been operating on a hotel and hospital laundry idea and finds that it is very possible to find out only later that "laundry" is not an industry covered by the managerial description of SSI. Date Another businessperson in Kalyan (West Bengal) created an Ayurvedic formulation but found that the DIC office did not possess a specialist to approve the initiative; the item needed to be promoted as candy! It is clear from these instances that the businessperson must also be sure of the administrative and legal matters related to the project. This entails selecting the type of business structure, obtaining registrations, and securing approvals from different authorities.

1.3 Establish Terms of Reference

Once the solution is accepted and the funding established, the project is created. The mandate outlines the vision, goals, scope, and outcomes of the project. It also encompasses the organizational framework (positions and duties) and a summary plan of the actions, resources and funding needed to implement the project. Finally, all risks, problems, design assumptions and limitations are listed.

What constitutes a Terms of Reference?

A Project Charter outlines the objective of the project, the way the project will be organized and how it will be effectively implemented. The Project Charter describes the project vision, aims, boundaries and outcomes, along with the Stakeholders, roles and responsibilities. The Project Charter is also known as a or "Project Definition Report".

When do we utilize a Terms of Reference?

These instructions assist you in determining the boundaries of your project. Writing the terms of reference is often one of the most challenging stages of the project life cycle, because it defines the parameters within which the project needs to be executed. It defines the vision, goals, range and implementation of the project, giving the team clear boundaries within which, the project must be carried out.

These task conditions will help you.

- Determine the vision and goals of the project.
- Define the overall scope of the project.
- List all critical deliverables of the project.
- Notify clients and project stakeholders.
- List the key roles and their responsibilities.
- Create an organizational structure for the project.
- Document the overall implementation plan.
- List all risks, issues and assumptions.

1.4 Appoint Project Team

At this stage, the boundaries of the project are clearly outlined and the project team is prepared to allocate according to the specifications or categories of the project. Although the project manager may be designated at any phase of the project, he must be appointed before the project group is created. The project leader records an elaborate job description for every project position and assigns human resources to each position based on their applicable abilities, knowledge and experience. When the team is "fully equipped", the project office is prepared to be set up.

1.5 Setup a Project Office

A project office is a tangible space/environment where a team works and plans work. While it is common for a project office to maintain a single central office, it is feasible to create a "virtual office" environment where project team members are situated in various regions around the globe. No matter the location, a successful project office configuration/environment comprises the following elements:

- Place (whether tangible or online)
- Communication (telephone, computer network, email, internet connectivity, file storage, database storage, and backup resources)

- Documentation (approach, procedures, form and recorded)
- Equipment (for bookkeeping, project scheduling, and risk assessment)

1.6 Perform Phase Review

At the conclusion of the initial face, a phase assessment is conducted. It is essentially a verification point to confirm that the project has accomplished the set goal as planned.

What is a Project Review?

A project audit is an evaluation of the condition of the project at a certain point in time. For the first time during the duration Of the project, the evaluation of the project is conducted at the conclusion of the initial phase of the project, called "Initiation". During this project review, it is decided whether the team has achieved its goals and approval is given to move to the next project phase, which is the "Planning" phase. Conducting a project management evaluation at the conclusion of every phase is essential for the project's success because it enables the project sponsor to track the project's advancement and ensure that it is moving smoothly through each project phase.

When do we complete a Project Review?

As soon as the project group thinks it has finished a specific stage of the project, the project must be reviewed. Throughout the project's life cycle, there are typically a minimum of three project assessments: at the beginning, during the planning stage and upon completion of the implementation phase.

2. Project Planning

Once the project's scope is established in the task, the project progresses to the detailed planning phase. It contains:

- Project Plan (describing the activities, tasks, dependencies, and timelines)
- Resource Plan (detailing the labor, equipment, and materials needed)
- Financial Plan (specifying the costs of labor, equipment, and materials)
- Quality Plan (outlining quality objectives, assurance, and control measures)
- Risk Plan (pointing out potential risks and the measures taken to reduce them)
- Acceptance Plan (detailing the criteria that must be met for customer acceptance)
- Communications Plan (specifying the information needed to update stakeholders)
- Procurement Plan (specifying the products to be acquired from external vendors).

At this stage, the project has been meticulously planned and is set to be carried out.

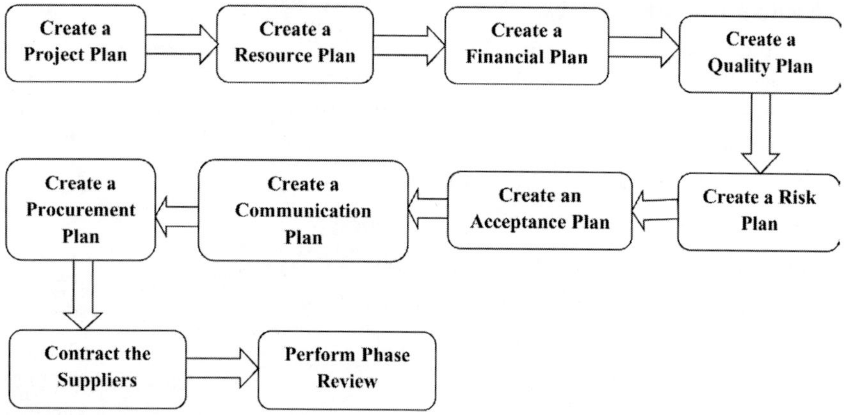

Figure 2.5: Phases of Project Planning

2.1 Create a Project Plan

The initial step is to record the project plan. A Work Breakdown Structure (WBS) is defined that includes the hierarchical phase, activities and tasks of the initiative. Once the WBS is agreed upon, an estimate of the work needed to finish the activities and tasks is arranged. Activities and tasks are assigned, resources are designated and a comprehensive project timeline is created. This project schedule becomes the project manager's main instrument for monitoring project advancement. The project plan is the crucial document of the project because it provides the project manager a road map to move forward and informs him along the way whether he is on the right track. By utilizing this project plan template, you are able to develop a thorough project management plan for your project today.

What is a Project Plan?

The project plan outlines the steps, functions, and tasks required for executing the project. It also displays the timelines, resources, and milestones essential for carrying out the project. Using this project plan template, you can swiftly and effortlessly develop a comprehensive project management plan for your project, since it already enumerates the frequently utilized tasks necessary to finish projects from beginning to end.

When is a Project Plan Template utilized?

The project template gets completed each time we intend to initiate a new project. The overall the project plan is typically created during the initial phase, while the detailed project plan is developed later in the design phase. The project plan is continually referred to during the project. Each day, the project manager assesses the real progress in comparison to what is documented in the project plan to ensure they remain on course. Consequently, the project plan is the most crucial instrument a manager can possess for successful implement projects.

2.2 Develop a Resource Plan

Immediately following the creation of the project plan, it is essential to distribute the necessary resources to finalize each function and task of the project plan. Although overall resource groups can presently be designated in the project plan, an exhaustive resource evaluation is necessary to determine:

- Types of labor required for the project
- Roles and key responsibilities for each labor type
- Number of people required to fill each role
- Items of equipment to be used and their purposes
- Types and quantities of equipment needed
- Total amount of materials needed

This Resource Plan will also help in

- Plan the dates when these resources will be used or consumed
- Determine the amount of resource required for a project activity
- Develop a comprehensive resource allocation timetable

What is a Resource Plan?

The resource plan outlines the amount of resources required to finish the project. A properly documented resource plan defines the exact amounts of labour, equipment and materials needed to complete your project.

When do we use a Resource Plan?

The resource plan is created during the resource planning stage of the project. Each project resource manager must prepare a comprehensive resource plan to guarantee that all resources required to finalize the project are recognized. By implementing the right planning tools, it will help you budget and forecast project costs.

2.3 Develop a Financial Plan

Similar to the resource plan, a financial plan is created, where it is explained how much money is needed for every phase of the project. The overall expense of labor, equipment, and materials is calculated and a cost schedule is defined, allowing the project manager to comprehend estimated expenses compared to real expenses during the project. Creating a detailed a financial plan holds significant importance as the project's success hinges on its completion within the "time, cost and quality estimates" allocated for that project.

What is a Financial Plan?

The financial strategy establishes the funding for the project.(i.e., money), which is essential to reach specific objectives. The financial plan outlines all the various kinds of expenses linked to the project (labour, equipment, material and administrative costs) and the estimated value of each cost. The financial plan additionally includes an overview of the overall expenses for the complete project, and these total costs become the project budget. As part of financial planning, a schedule is drawn up that lists the amount of money needed at every phase of the project.

When do we use a Financial Plan?

If we need to ask for money, we require a dependable financial strategy that illustrates how it will be utilized. For a project manager, acquiring project funding is among the most essential responsibilities of a project. Therefore, sound financial planning principles must be followed to ensure a positive outcome. This financial plan template allows us to create a detailed financial plan for our project. This will help us obtain the necessary project funding to successfully deliver our project on time.

This Financial Plan template will assist us in recognizing the:

- Types of labor expenses incurred throughout the project
- Machinery required to execute the project
- Various materials necessary for the project
- Cost per unit of labor, machinery, and materials
- Other types of costs, such as administration
- Number of contingencies required

We can then use the Financial Plan template to create a budget by:

- Calculate the total cost of completing the project
- Determine the total cost of each project activity
- Create a cost plan

2.4 Quality Plan

Quality planning is an important part of all projects. This allows you to agree on quality goals with the client. This allows you to oversee and regulate the quality standard generated by the project to confirm that the set quality objectives are met. Using this quality plan template, you can set quality goals and ensure that your project delivers results that meet your customers' needs, ensuring your success.

What is a Quality Plan?

A quality plan allows you to plan all the necessary actions to guarantee that your project fulfils the requirements of your client. It is composed of two sections; The quality control plan lists the necessary independent audits and the quality control plan lists the internal audits necessary to meet your quality objectives. Utilizing quality assurance and quality assurance methods, you can create a comprehensive quality control strategy for your project.

- Meeting the clients' quality expectations is essential for the project's success. The quality plan is recorded to guarantee that quality expectations are explicitly outlined and reasonable. Quality Plan:
- Clarifies the meaning of quality for this project.

- Define clear and specific goals for each task. Each quality objective contains criteria and standards that need to be fulfilled to satisfy the customer's expectations.
- Creates an action plan that ensures the client achieves quality goals (i.e., quality plan).
- Recognizes the methods employed to manage the real quality level of every output as it is created (i. e. , quality control plan).

When do we use a Quality Plan?

Formulating a quality plan is essential if we want to give the customer confidence that we are producing a solution that meets their needs. The quality plan tells us everything we intend to do to ensure the quality of our solution. The first part defines the quality objectives. The second part presents the quality assurance plan. And the third part defines the quality control plan. Using this model, we can create a quality management plan that gives our client great confidence in success.

2.5 Develop a Risk Plan

Predictable risks are then documented in a risk strategy, and a collection of measures must be prepared to prevent the occurrence of each risk and lessen the effect of the risk when it occurs. Creating a well-defined risk plan is a vital step during the planning stage, since it is essential to mitigate all significant project risks prior to the project execution phase.

Creating a risk management a plan is an essential stage for every project, as it helps reduce the probability of a risk happening. Regardless of the kind of risk, you can use this model to implement processes and procedures to lessen the likelihood of the risk, helping you to successfully deliver the project.

What is a Risk Plan?

A risk management plan assists us in predicting potential risks, identify measures to prevent them and reduce their impact when they occur. A risk management plan is prepared as part of the risk planning process. It lists all foreseeable risks, their order and priority, preventive and possible actions and their monitoring process. This risk plan template helps us go through these steps quickly and easily.

When do we use a Risk Plan?

A risk plan ought to be utilized whenever it is necessary to manage risks with great care. For example, during project initiation, a risk management plan is created that recognizes and manages the risk associated with project implementation. The risk plan is often referenced throughout the project so that all risks can be mitigated as quickly as possible. A risk plan model helps us identify and manage risks, which increases our chances of success.

This Risk Planning template will help to

- Determine the risks in our project
- Classify and prioritize each risk
- Assess the likelihood of the risks happening
- Evaluate the effect of the risk occurrence on the project

We can then use this Risk Plan template to:

- Determine precautionary steps to avert the risk from arising
- Identify potential actions to lessen the effects if the risk materializes.
- Plan these measures in a reasonable time frame
- Monitor the progress of every risk during the project

2.6 Develop an Acceptance Plan

The essential factor for a successful project is getting the customer's approval that every product manufactured satisfies (or surpasses) their criteria. The acceptance plan is created to clarify the acceptance plan, based on which criteria the acceptance of every delivery by the customer is assessed. The acceptance plan outlines the standards for securing customer approval, along with a timeline for acceptance evaluations during which customer approval is sought, and a summary of the procedure through which each delivery is received by the customer.

What is an Acceptance Plan?

An acceptance plan (also referred to as an "acceptance test plan") is a timeline of activities needed to get customers to accept that your product is satisfactory. However, it is more than just a to-do list. An acceptance plan is actually a contract between you and the client that states the acceptance tasks that will be finished at the conclusion of the project for ultimate approval. The admission strategy consists of a compilation of outcomes, admission testing activities, criteria and standards to be fulfilled and a strategy for their execution.

When do we use an Acceptance Plan?

We need to create an approval plan every time there is a set of deliverables that require customer approval to complete. If something needs to receive approval from the customer, you should agree in advance what steps will be acquired to receive authorization after the deliveries are completed. Preparing the approval strategy at the onset of the project saves us time and effort in the final phase, because the acceptance testing activities have already been done by the client.

2.7 Develop a Communication Plan

Before the implementation phase, it is also essential to determine how each stakeholder will be informed about the advancement of the project. The correspondence plan defines the categories of information to be communicated, the techniques of communication information with stakeholders, the rate of distribution, and the duties of each project team member for regularly sharing information with stakeholders. This communication plan helps deliver the right information to the right people at the right time.

What is a Communication Plan?

A communications plan (or communications strategy) outlines how you intend to convey the right messages to the right people at the right time. The communication plan describes the communication objectives, stakeholders and strategies, activities and timelines. A communications strategy will assist you in keeping everyone updated so you can provide a consistent message to your target audience.

When do we use a Communication Plan?

If we have many employees, external suppliers, customers and stakeholders to communicate with, we should formally record our communications in a communications plan. A clear communication plan is essential to the achievement of an organization. It is also vital to the success of initiatives because it ensures that all employees and stakeholders are properly informed about project progress. The best time to plan communication is during the initiation stage of the project. This ensures that our communication plan includes the tasks necessary for successful communication during the project life span.

We can then use this Communication Plan for

- Monitoring completed communication events
- Receiving feedback from communication events
- Improving communication processes.

2.8 Develop a Procurement Plan

The sustaining activity of the planning phase aims to determine the project components that will be purchased from suppliers outside the project. The procurement strategy offers a comprehensive outline of the items (i. e. , goods and services) that will be acquired from vendors, the reasons for obtaining each item externally rather than internally by the business, and the timeframe for procurement. It also covers the method for choosing favored suppliers ("bidding process") and the method for precisely ordering and receiving the acquired items ("procurement process").

What is a Procurement Plan?

A procurement plan identifies the products and services you will purchase from outside vendors. A good procurement plan goes further by outlining the process you will follow to contract with these vendors. Whether you are embarking on a project procurement plan or an organizational procurement plan, the steps are the same. First, determine the items you need to buy. Next, outline the steps for obtaining these items. And finally plan your delivery time.

When do we use a Procurement Plan?

You should create a sourcing plan whenever you want to purchase from a supplier. Using a sourcing plan template, you can define your sourcing requirements, identify potential suppliers, enter into contracts with those suppliers, and manage them to ensure delivery. Project procurement planning is essential to the success of any project.

This Procurement Plan helps in

- Identify purchasing needs
- Identify all items to be purchased
- Create a solid financial basis for their purchase.

It also helps to

- Enumerate all the activities associated with product procurement
- Plan for these tasks by allocating time and resources
- Create a robust project procurement process for the company.

2.9 Contract Suppliers

Even though external suppliers may be engaged at any point during the project, they are typically appointed after the project plan has been recorded, but prior to the project execution phase. Only at this point does the project manager possess a clear understanding of the supplier's role and delivery expectations. A formal bidding procedure is started to identify interested suppliers and choose a favored supplier to fulfill the project's procurement requirements. The RFP process includes the development of a Statement of Work, Request for Information, and Request for Proposal to collect scientific data from every possible supplier in order to choose a preferred supplier. After a preferred supplier has been chosen, a supplier contract is executed to provide the necessary deliverables.

Statement of Work

What does a Statement of Work mean?

A Statement of Work (SOW), outlines what you require from an external vendor. It is a declaration of the tasks that a vendor is to execute. The statement of work also describes the materials and equipment to be provided within a specified time frame. This statement of work template saves time because all sections are pre-filled.

When do we use a Statement of Work?

Whenever we need to request work from an external vendor, we have to set up a Statement of Work. It helps us clarify what we expect from our vendor and the timeframe within which we have to achieve it. Writing a statement of work is easy using this comprehensive template.

This Statement of Work template helps us by:

- Identify the type of supplier we want to appoint
- Describe the materials and equipment we need
- Specify the deliverables the supplier will provide
- State our payment terms and conditions

Request for Information

What does a Request for Information entail?

A request for information (or RFI) is a document issued to potential suppliers to allow them to participate in the bidding process. It is an inquiry for details from the supplier to assist us in determining if we should engage them or not. By using this request for information form, we can complete the request for information process quickly and easily.

When do we use a Request for Information?

We must make a Request for Information each time we want to use an external supplier for our business. The document will explain the supplier and the information we need to help us select our supplier. This Request for Information template includes all the sections, content, and examples we need to record a Request for Information today.

This Request for Information template helps with:

- Define our supplier requirements
- Inform suppliers about our procurement process
- Specify rules and deadlines for interacting with suppliers
- Select suppliers for our business.

Request for Proposal

What is a Request for Proposal?

A Request for Proposal (RFP) is a document sent to suppliers to help them provide the details they require to make a decision on their preferred supplier. Using this RFP template, you can quickly create a request for proposal for your business. It includes all the relevant content you must develop a request for proposal.

When do we use a Request for Proposal?

We need to create a request for proposal whenever we want to request a proposal from external suppliers. The Request for Proposal describes the nature of the proposal we expect from them and the delivery schedule. The Request for Proposal also includes the key terms and conditions that we intend to include in the agreement with the supplier. This RFP template will save us time when creating an RFP now.

This Request for Proposal template will help us to

- Inform suppliers of our purchasing needs
- Inform suppliers of the information we need from them
- Make decisions about our preferred suppliers.

Supplier Contract

What is a Supplier Contract?

A supplier contract, or "supply contract," is an agreement between a company and an external supplier to provide a defined set of products and services. A supplier contract is a legal agreement and serves as a basis for measuring supplier performance. In addition to outlining the items to be provided, a supply contract also outlines the timelines, responsibilities, pricing, and payment conditions necessary to handle the relationship. By entering into a supply contract, you can maximize the benefits of your supplier relationship.

When do we use a Supplier Contract?

Supplier contracts should be used whenever we need to acquire goods or services from an outside vendor. By documenting the supply contract, our external suppliers will take the relationship more seriously. The supply contract defines the delivery stages and therefore the payment criteria. So, by using a supply contract, it will be easier to manage our suppliers today.

This template helps us document a supply contract that specifies the

- Deliverables provided by the supplier
- Training, documentation and support required
- Responsibilities of both parties
- Performance criteria and review process
- Price list and payment process
- Contract conditions.

2.10 Perform Phase Review

At the conclusion of the planning phase, a review of the phase is performed. It serves as a checkpoint to verify that the project has met the intended goals.

What is a Project Phase Review?

A review of the project phase takes place at the conclusion of every project phase. In the course of this project management review, the reviewer completes a phase review document that outlines the advancement of the project to date and recommends whether to proceed to the subsequent stage of the project. If approved, the subsequent stage of the project can be executed.

When do we use a Project Phase Review?

A project phase review should be carried out at the conclusion of each project phase. The project review can be conducted by the team leader or by someone independent of the project. Throughout the project management review, all risks and issues should also be recorded. A Project Phase Review form should then be completed, documenting the results of the review for approval.

This Project Phase Review Form states whether the

- Project on time and on budget
- Deliverables produced and approved
- Risks controlled and mitigated
- Issues resolved
- Project is on track.

The Phase Review Form helps us to

- Document our project review results
- Clearly communicate our project progress to the sponsor
- List any risks or issues that have impacted the project
- Show our sponsor the deliverables to date
- Seek authorization to progress to the subsequent phase of the project.

3. Execution

The execution stage is generally the most extended phase of the project (in terms of duration). This is the stage during which the deliverables are constructed and shown to the customer for approval. To guarantee that the customer's needs are fulfilled, the project manager oversees and regulates the activities, resources, and expenses necessary to create each deliverable throughout the execution phase. Various management processes are also applied to confirm that the project is advancing according to plan.

3.1 Build Deliverables

This phase requires the physical construction of every deliverable for client approval. The actual activities performed to build each deliverable will differ based on the kind of project, in which case the deliverables may be constructed in increments. Regardless Of the technique utilized to create each deliverable, thorough oversight and management must be applied to guarantee that the end deliverable satisfies the requirements or anticipations of the final customer.

3.2 Monitor and Control

The project team truly creates the final product, while the project manager conducts a sequence of management processes to oversee and regulate the activities carried out. A summary of each business process is as follows:

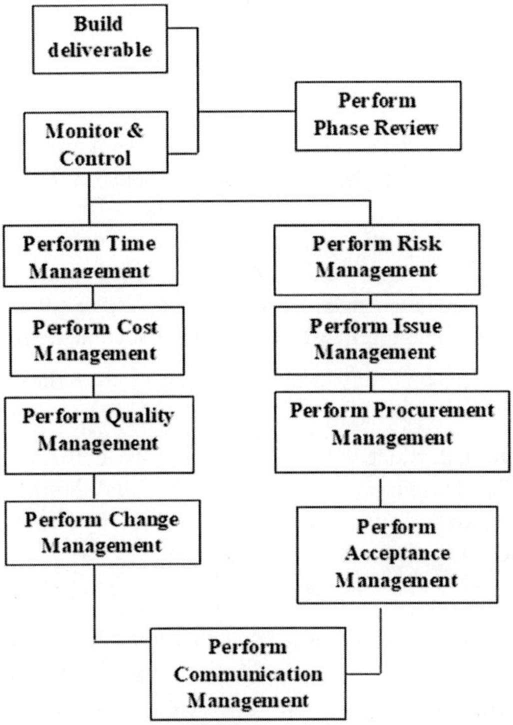

Figure 2.6: Phases of Monitor and Control

3.3 Time Management

Time management refers to the act of documenting the time an employee spends performing project tasks as part of a project. As time is a limited and critical resource in projects, It is crucial to log the time utilized by each team member on a timesheet to enable the project manager to manage the level of resources allocated to a particular activity. A specific activity timesheet log provides a summary of an employee's existing time dedicated to a particular project task and assists in revising the project plan.

3.4 Cost Management

Cost management is the procedure through which the costs (or expenses) incurred on a project are officially recognized, approved, and paid for by an authorized body. Cost forms are filled out for every expense category of the project, such as labour, equipment, machinery, individual components, and materials. The cost forms are authorized by the project manager and documented in the expense records for auditing reasons.

3.5 Quality Management

Quality is described as "the degree to which the final product conforms to customer requirements". Quality management is the procedure of ensuring and controlling the standard of project outputs, utilizing quality assurance and quality control methods.

Regular quality assessments are performed and results are recorded in a quality register.

3.6 Change Management

Change management is the procedure through which modifications to the project scope, deliverables, process, or resources are recognized, formally reviewed and authorized prior to execution. A crucial element of the project manager's responsibilities is to effectively handle alterations within the project. To do this, the organization must understand the operational and system elements that require change, document the advantages and expenses of executing the change, and create an organized strategy for carrying out the change. To officially seek a change, it is often necessary to complete a change request form. Details of the change request may then be documented in a change log.

3.7 Risk Management

Risk management is the procedure through which project risks (e.g. scope, deliverables, timelines or resources) are officially recognized, measured, and handled throughout the project. Project risks can be recognized at any point during the project by filling out a risk form and documenting the pertinent risk information in a risk register.

3.8 Issue Management

Issue management involves the official handling of issues that are presently impacting the project's capacity to deliver the necessary deliverables. After completing the issue form (and recording the details in the issue log), the project manager will evaluate each issue and implement a sequence of steps to address the problem at hand.

3.9 Procurement Management

Purchasing management is the procedure of purchasing products from outside vendors. To ask for a supplier to deliver a product, the order needs to receive approval from the project manager and be forwarded to the supplier for confirmation. The status of the purchase is subsequently monitored through a purchase log until the product is delivered and accepted by the project team.

3.10 Acceptance Management

Acceptance management is the process through which the outputs generated by a project are evaluated and authorized by the client as satisfying their particular requirements. Therefore, in order to request customer acceptance of the deliverable, an acceptance form is filled out. The acceptance form outlines the standards used to produce the deliverable and the degree of satisfaction for each listed criterion.

3.11 Communications Management

Communications management refers to the process through which official communication messages are recognized, developed, assessed, and conveyed within a project. The most prevalent approach for communicating project status is through project status reports. Each piece of communication disseminated to project stakeholders are documented in a communication log.

3.12 Perform Phase Review

At the conclusion of the implementation phase, a phase review is conducted. It is essentially a checkpoint to confirm that the project has reached the planned objectives.

4. Closure

Once all project outputs are finalized and approved by the client, a successful project will have met its objectives and be prepared for formal closure. The closure of the project is the concluding stage of the project and must be carried out officially for the client to completely understand the business advantages the project provides.

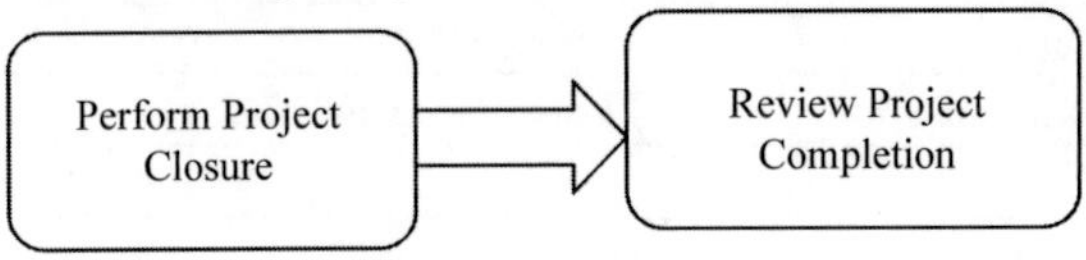

Figure 2.7: Closure

4.1 Perform Project Closure

Project closure consists of executing a set of tasks to conclude the project, which includes:

- Evaluate whether the project completion standards have been met.
- Identify outstanding elements (activities, risks, or issues).
- Develop a handover plan to transfer the product to the customer environment.
- Enumerate the actions needed to move documentation, terminate the contract with the supplier, and free up project resources to the company.
- Communicate the closure to all stakeholders and interested parties.

The project completion report is presented to the client and/or project sponsor for approval. The project manager is then accountable for finishing each task specified in the project completion report on schedule and within financial limits. The project is not closed until all tasks listed in the project completion report have been finalized.

4.2 Review Project Completion

The last task completed in a project involves assessing the overall success of the project by an external resource. Success is defined by the degree to which the project has reached the identified objectives and by its compliance with the managerial processes outlined throughout the planning stage. To assess productivity, several questions are asked.

Examples

- Does the project deliver the advantages outlined in the business case?
- Does it fulfill the goals outlined in the terms of reference?
- Does it function in accordance with the terms of reference?

- Does the deliverable comply with the standards specified in the quality plan?
- Is it provided within the timeline outlined in the project plan?
- Is the project being carried out within the financial resources allocated in the financial plan?

To assess compliance, an evaluation of the degree of adherence of project activities to the management processes outlined in the quality plan will be conducted. The aforementioned outcomes, significant accomplishments, and insights gained are documented in an Implementation Review Report and presented to the project sponsor for endorsement.

Integration	**Scope**	**Time**
Cost	Quality	Procurement
Human resources	Communications	Risk management
Stakeholder management		

All management indeed deals with these matters. However, project management introduces a distinct emphasis influenced by the objectives, resources, and timeline of each project. The significance of that emphasis is demonstrated by the swift, global expansion of project management:

- As an acknowledged and strategic capability of the organization
- As a topic for training and education
- As a career path

Project management is a systematic method for organizing and directing project activities from beginning to end. As per the Project Management Institute, the processes are directed through five phases: initiation, planning, execution, monitoring, and closing. Project management can be utilized for virtually any kind of project and is commonly employed to manage the intricate processes of software development projects.

The systems development life cycle (SDLC) is one example of a framework for steering the project management process from an initial feasibility study through maintenance of the completed application. Various SDLC approaches include the waterfall model, which was the original SDLC method; rapid application development (RAD); joint application development (JAD); the fountain model; the spiral model; build and fix; and synchronize-and-stabilize. A number of charting methods, such as the Gantt chart and PERT charthave been developed as tools to create a graphic representation of a project plan and its current status.

Project-driven Organizations

To compete in an increasingly competitive and dynamic global marketplace, organizations are investing in opportunities to transform from functionally oriented businesses to project-oriented businesses. The project-based organization refers to the processes that can help organizations transform, processes that include implementing incremental change, adopting a systems focus, developing organizational maturity, conducting structured assessments, and quantifying return on investment. It identifies each process, its purpose, and the detailed methods that can help organizations implement each process. It also provides insight into the challenges organizations face when transforming from a functionally oriented business to a project-oriented business.

What Is the Difference Between Project-driven And Non-project-driven Organizations?

There are several **differences between project-driven and non-project-driven organizations**, including:

- Project management in project-based organizations is mature and respected. In contrast, in non-project-oriented organizations, project management is still in its infancy and is often viewed with skepticism.
- Project-based organizations derive most of their revenue from projects, while non-project organizations derive their revenue primarily from production activities.
- Project managers hold the responsibility for profits and losses in project-based organizations. In non-project organizations, the accountability for gains and deficits is ambiguous.
- Project-based organizations adopt a purely project-based or matrix organizational structure. Non-project organizations often adopt a functional organizational structure.
- Project-oriented companies have flexible career paths that allow for rapid advancement to higher positions. Non-project-oriented companies have traditional career paths that make it difficult to move up in the company. Often, you have to wait until your manager is fired, resigns, retires, or dies to move up in the company and take on a better position.

  Examples of industries where project-driven organizations are predominant include:

  - Construction
  - Aerospace
  - Research
  - Value added

  Examples of industries where non-project-driven organizations are predominant include:

  - Manufacturing
  - Services
  - Farming
  - Natural resources
  - Agriculture

IT (Information Technology) organizations (which are part of the service sector) are considered hybrids, where parts of such organizations are considered project-based (such as developing new software) while other parts are considered project-based. non-project-oriented (e.g. support applications).

Functional Organization Structure

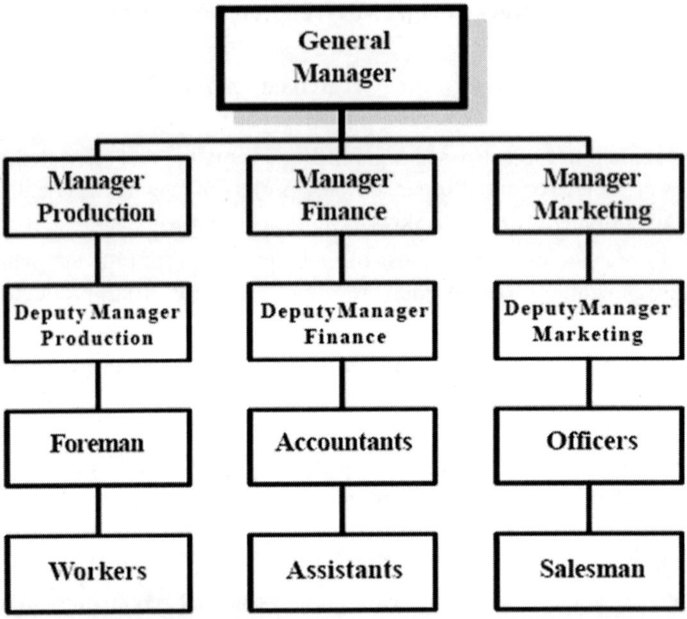

Figure 2.8: Functional structure of organization

A functional organizational framework is a hierarchical organizational structure where people are grouped in accordance with their field of expertise. These individuals are overseen by a CEO who possesses expertise in the same domain. This expertise helps him to effectively use the skills of the employees, which ultimately helps the organizations to achieve their business goals.

In such an organizational structure, people are classified according to the role they fulfill within the organization. The organizational diagram of the functional organizational structure displays the CEO, vice president, finance department, sales department, customer service, administration, etc.

Each department has its own department head who is accountable for the functioning of his department. It assists the organization in overseeing the quality and consistency of performance. These various departments are sometimes called silos. This means that the system is vertical and disconnected. Communication goes through department heads to upper management.

Here, all authority (i.e., budget allocation, resource allocation, decision making etc.) rests with the functional manager. Typically, a project manager role is not present within this kind of organizational framework. Even if this position exists, the responsibility of the project manager is quite restricted and necessitates the consent of the functional manager to fulfill the requirements. The project manager may have the rank of coordinator or supervisor. A functional organizational structure is suitable for an organization that engages in continuous operations and produces standard products or goods, such as an industry.

Advantages of the Functional Organization Structure

The following are a few benefits of the functional organization structure:

- Employees are grouped according to their knowledge and skills, which helps to achieve the highest performance.
- The staff is very professional. Efficiency is achieved because they possess experience in the identical role and they perform exceptionally well.
- Their duties and obligations are set, which facilitates accountability for work.
- Hierarchy is very clear and employees do not have to report to several managers. Each employee reports to their functional manager, which reduces the number of communication channels.
- There is no duplication of work, because each department and each employee has a specific work responsibility.
- Employees feel safe and therefore work well without fear.
- Because the workplace is safe, employees tend to be loyal to the organization.
- Employees have a clear career path of growth.
- Cooperation and communication in the department is excellent.
- Employees do not report to multiple supervisors. Each employee reports directly to their functional manager. This reduces communication channels.
- Employees tend to gain knowledge from one another because employees with the same specialty (or skill) are grouped together.

Disadvantages of the Functional Organization Structure

The following are a few disadvantages of the functional organization structure:

- Employees may feel bored and lose enthusiasm for work due to monotonous, repetitive work.
- If the performance evaluation system is not properly managed, conflicts may arise. For example, an employee may feel depressed when a lower performing employee is promoted.
- A highly educated worker pays more.
- Expert way of thinking is self-centered. A functional manager pays more attention to his department; he usually does not care about other subjects.
- Interdisciplinary communication is weak, which impairs interdisciplinary coordination. This reduces flexibility and innovation. In addition, there is no teamwork between different departments.

- Employees may have little concern and/or knowledge of events outside the department. This creates obstacles in communication and cooperation.
- The operational structure is rigid, which makes it difficult and slow to adapt to changes.
- Delays in decision-making frequently occur because of the bureaucratic hierarchy.
- Generally, a functional manager makes decisions autocratically without consulting his team members. This may not always work in favor of the organization.
- As an organization grows, functional areas can become difficult to manage because of their size. Each department can begin to behave like a small business with its own region, culture and management style.
- Departmental goals can distract from functional departments and focus on them instead of the organization's goal.
- No one "controls" a specific project, the project is managed by several functional managers, this can lead to the following:
- The client cannot know what the real status of the project is, because no one can tell him the whole picture. Every functional manager is able to report on his segment of the project.
- Conflicts among functional managers can dramatically affect project progress.
- Workers usually spend a significant portion of their time doing nothing, sometimes literally begging for work.

Project Report

A project report is a document that outlines the overall perspective of a suggested business. The project report gives a description of the project proposal to clarify the future possibilities of the suggested plan/activity. A project report is a written document associated with any investment. It includes the information upon which the project was assessed and deemed viable. It contains details on economic, technical, financial, management, and production factors. It assists the entrepreneur in understanding the interests and aids in obtaining loans from banks or financial institutions.

The project report includes comprehensive details on the required land and buildings, annual production capacity, production process, machinery and equipment along with their prices and specifications, raw material needs, electricity and water requirements, labor requirements, marketing costs. Project, production, financial assessments and financial feasibility of the project.

Contents of a Project Report

Below are the details of a project report.

1. **General Information:** The project report needs to encompass details regarding the industry associated with the project. It should present information about the industry's historical background, present condition, challenges, and future

outlook. It must include details about the product intended for production and the rationale behind the product selection if the suggested company is a manufacturing entity. It should outline the demand for the product in local, national, and global markets. It needs to explicitly articulate the business opportunities and clarify the justifications for initiating the business.

2. **Executive Summary:** The project report ought to discuss the business goals and the approaches through which the organization can attain success. The project report is required to provide an overview of the company regarding its capital, operations, operational methods, and business implementation. It should address the assumptions and risks related to the company overall.
3. **Organization Summary:** The organizational structure and model the unit should be referenced in the project report. It should be observed if the foundation of the ownership right is a limited company, a company or a joint stock company. It is required to present details regarding the driver's biography, including financial stability. The project report must include the name, address, age, qualifications, and experience of the owners or implementers of the proposed company.
4. **Project Description:** A concise summary of the project should be provided and should include information regarding the following:
   - Position of the site
   - Raw material needs
   - Aim of production
   - Space needed for the work shed
   - Power requirements
   - Fuel requirements
   - Water requirements
   - Requirements for skilled and unskilled labor employment
   - Technology chosen for the project
   - Production process
   - Estimated output quantities, per-unit pricing
   - Waste management facilities necessary

   If the business focuses on services, it must specify the kinds of services offered to clients. It should describe the manner of delivering services to customers thoroughly.
5. **Marketing Plan:** The project report should explicitly express the anticipated overall demand for the product. It should specify the price at which the product can be available in the market. The tactics employed to seize the market should also be included. If after-sales service is provided, this must likewise be noted in the project. It should outline the method of distributing the product from the production unit to the market. The following must be included in the project report:

- Category of clients,
- Target markets,
- Nature of market,
- Market segmentation,
- Prospective future of the market,
- Sales objectives,
- Marketing expenses of the project,
- Market portion of the suggested enterprise,
- Interest in the product within the local, national, and international markets,
- It should specify possible users of the products and distribution methods that will be utilized for delivering the product.

6. **Capital Structure and operating cost:** The project report should outline the overall capital needs of the project. It should indicate the financing sources; it should also state the amount of the capital from owners and borrowed funds. The project must detail the working capital necessities along with the source of procurement. The appraisal of the total project cost needs to be categorized into land, building construction and earthworks, equipment and machinery, various fixed assets, initial and pre-operational costs and working capital. The proposed financial the composition of the project must specify the anticipated sources and conditions of equity and debt financing. This section should also include the operating expenses.
7. **Management Plan:** The project report must include the subsequent information.
   - Business experience of the business promoters,
   - Information regarding the management team,
   - Tasks and obligations of team members,
   - Present staffing requirements of the organization,
   - Techniques for overseeing the business,
   - Strategies for recruiting and educating staff,
   - Policies and programmes of the management.
8. **Financial Aspects:** To evaluate the business's profitability, the anticipated profit statement and balance sheet need to be included in the project report. It should display the forecasted sales revenue, production expenses, gross profit, and net profit expected to be generated by the proposed organization. Furthermore, along with the aforementioned, a projected balance sheet, cash flow statement, and cash flow must be created annually and cover a minimum period of 3-5 years.

   The income statement and cash flow forecast are required to feature a three-year summary, monthly information for the first year, and quarterly data for the second and third years. The project report should specify the break-even point and return on invested capital. The accounting system and inventory control system that are utilized are typically covered in this section of the project report. The project report must clarify if the company is economically and financially viable.

9. **Technical Aspects:** The project report contains details regarding the technology and technical elements of the project. This encompasses details about the technology selected for the project, the manufacturing process, machine capacity, pollution management equipment, and so on.
10. **Project Implementation:** Each suggested business unit is required to create a project timeline. It should specify the duration in which activities concerning the formation of the company can occur. Execution plans illustrate the timeframe for project development and finalization.
11. **Social responsibility:** The suggested units will obtain their support from the community. Therefore, its input to the community regarding employment, income, exports, and infrastructure. The project report must showcase the business output.

Project Appraisal

The assessment of the viability or feasibility of the suggested project by financial institutions is referred to as project appraisal. An alternate definition of project appraisal is: "A lender's evaluation of the viability of a project during the repayment period is known as project appraisal".

Four primary standards in project evaluation

There are four primary criteria that must be examined in project evaluation to determine its viability. They are:

- Technical viability
- Economic viability
- Financial feasibility
- Managerial viability

Technical feasibility of projects

- To determine if the different factors of production are accessible
- Appropriate placement of the project
- Implementing suitable technology
- Offering adequate training to workforce
- Installation of plant and equipment
- Technical know-how
- Appropriate plant arrangement
- Approval for pollution from the pollution control authority

- Environmental approval under the Environmental Protection Act
- Safeguarding under the Patent Rights and Trademark Act
- Waste disposal

Economic feasibility of projects

- Product market share.
- Product demand.
- The market is competitive.
- Product life cycle and product stage.
- The anticipated requirement for the product.
- Realization of social goals such as job creation, improvement of underdeveloped regions, etc.
- Scope of the item [Strength, Weakness Opportunity and Threat (SWOT Analysis)].

Financial feasibility of projects

- The financial stability of the project that relies on return on investment
- Different funding sources accessible and their associated costs
- Predicted cash inflow and outflow
- Expense of the project
- Profit margin
- Expense of manufacturing
- Prospective expansion of the initiative
- Total and net profits
- Future prospects

Managerial feasibility of projects

- Ability of the Board
- Expertise of the staff
- Technical competence
- Issues that may be faced in industrial relations
- Knowledge in the area

How to forecast demand for new products?

Forecasting the demand for new products requires special knowledge and techniques, because these are new products and there is no previous information about their sales. The method or techniques must be carefully adapted to the product. Joel Dean presents six possible approaches to predicting new products. These are the following:

1. **The Evolutionary approach in forecasting demand:** This approach is founded on the principle that the demand for a new product is only the growth and development of an existing product. This means that when looking for product demand, the demand conditions for the existing product must be taken into account. Examples: Color TVs for black and white TVs; Left-hand drive cars from right-hand drive cars etc. But this approach is useful only if the new product is very close to the old existing product.
2. **Substitute approach in forecasting demand:** By this the new product is analyzed as a substitute for the old existing product or service.
3. **Growth curve approach in forecasting demand:** Estimates of the growth rate and level of final demand of the new product is created based on several established product growth models. For example, the average sales of talcum powder give an idea of how new cosmetics are introduced to the market.
4. **Opinion Poll approach in forecasting demand:** Therefore, the demand for the new product is estimated by conducting direct surveys of the end consumers. This is done using a sample survey method. However, this is an extremely challenging procedure because of the issues related to sampling, researching the true intentions of consumers, etc.
5. **Sales Experience approach in forecasting demand:** According to the sales experience method, samples of new products should be offered in a sample market to anticipate demand. This is done through distribution channels such as department stores or cooperatives etc. or through direct mail. The total demand is estimated based on the sample market. But there are difficulties here in determining the immaturity of the sample market and the complete market.
6. **Vicarious approach in forecasting demand:** With the vicarious approach, customer reaction to a new product can be indirectly ascertained through specialized retailers who can assess customer needs, tastes and preferences. Retailers, who are the link between the manufacturer and the final consumer, get to know how customers receive the new product.

Which is the best method to forecast demand of product?

All of the above methods of forecasting demand for a new product are not mutually exclusive. According to Joel Dean, the evolutionary approach to demand forecasting is useful when the new product is simply an improvement on an old product. The growth curve method is applied only to a limited extent. The replacement approach seems to be very simple but very difficult to implement when one assumes the demand for new products in the market.

3

Network Methods and Project Scheduling

Construction of any building, airplane, poultry farm, dairy farm, agricultural business, even a new product are known as projects that require money and time, thus, all projects need to be finished in the least amount of time possible. All projects involve multiple activities, and all these actions need to be logically coordinated and successfully completed within the expected time frame. PERT and CPM are useful techniques for this purpose. PERT stands for Program Evaluation and Review Technique and CPM represent Critical Path Method. These techniques separate projects into various tasks and events and organize them logically in a diagram called a grid. Time estimates are obtained to conduct these different tasks and the total project time is decided based on these estimated times. With the help of these methods, the project can be completed in the given time and it also provides important information about the various activities related to the project and the project manager knows the importance of the various activities to which he pays more attention to reduce the time spent on the project.

Network analysis refers to several complex planning and management techniques projects.

The two most frequently used forms of network planning are

- Critical Path Method (CPM)
- Program Evaluation and Review Technique (PERT)

CPM was developed by Walker and Kelley at the Do Pont Company in 1957 and focused on the trade-off between project cost and total project completion time (eg, for certain activities, it might be feasible to lessen their completion time by spending more money. money - how does this affect the entire project completion time?). In CPM, activities are represented as a network of prioritized relationships that use activities to create nodes and are used for essentially repetitive work where activity time estimates can be predicted with relative certainty due to past experience.

PERT was created by the United States Navy in 1958 to plan and monitor the Polaris missile program with a focus on completing the program in the shortest possible time. In addition, Pert had the ability to handle uncertain activity completion times (e.g., the most probable finishing time for a given activity is 4 weeks, but it can vary from 3 weeks to 8 weeks). In PERT, an activity is represented as a prioritized relationship network using an arrow network construction and is used for non-recurring work (research and development) where time and cost estimates are usually very uncertain. This method uses probabilistic time estimates.

Both technologies were developed during the same period. Although both technologies were developed independently, they share the same goals. PERT and CPM are used not only in large projects worth rupees but also in projects worth several hundred rupees. The use of these technologies has now become expected in the planning, organizing, and overseeing different project tasks. CPM and PERT techniques are essentially network planning techniques used in the organization and administration of building, maintenance production, procurement and quality control, etc. They use technology to minimize problem areas such as production delays, delays and disruptions by identifying critical factors and coordinating different parts of the entire operation.

Details About Pert (ADM) and CPM (PDM)

The Program Evaluation and Review Technique (PERT) was created by the U. S. Navy in collaboration with Booz- Allen Hamilton and the Lockheed Corporation for the Polaris missile/submarine initiative in 1958. The Critical Path Method was first presented by DuPont, during that same timeframe.

In practice, PERT has primarily been employed for research and development projects, the specific kind of projects for which it was created, although its application is more prevalent on the "development" aspect of research and development than on the "research" aspect. CPM was intended for construction projects and has been widely accepted by the construction sector. (There are several exceptions to these broad statements. The Eli Lilly Company, for instance, applies CPM to its research initiatives.)

The utilization of PERT has declined significantly in recent times as a vast majority of project management software produces CPM networks. The two techniques are quite alike and are frequently integrated for instructional purposes.

Initially, PERT was specifically focused on the time component of projects and utilized probabilistic activity time estimates to assist in ascertaining the likelihood that a project could be finished by a specific date. Conversely, CPM employed deterministic activity time estimates and was created to manage both the timing and cost elements of a project, especially concerning time/cost trade-offs. In CPM, activities are capable of being "crashed" (accelerated) for an additional cost to hasten the completion time. Both methods identified a project's critical path with activities that couldn't be postponed, while also highlighting activities with slack (or float) that could be delayed without extending the project's completion time. Some authors maintain a strict distinction between PERT and CPM. We find this to be unwarranted. It is possible to estimate probabilistic CPM times, and one can "crash" PERT networks.

We may mention briefly that the essential activities in actual projects usually make up fewer than 10 percent of the overall activities. In the examples and simplified problems presented in this chapter, the essential activities represent a significantly larger share of the total because we utilize smaller networks to demonstrate the methods.

For nearly fifty years, PERT and CPM networks have been utilized in project management. Nevertheless, in 2005, the Project Management Institute (PMI) updated the names, referring to PERT as ADM (Arrow Diagram Method) and CPM as PDM

(Precedence Diagram Method) (Project Management Institute, 2005). Generally, we will simply refer to either method as a "network," but if we want to specifically mention a PERT (ADM) network where the activities are represented as arrows on the network, we will label it as an AOA (activity-on-arrow) network, and if we mention a CPM (PDM) network where the activities are depicted as nodes, we will identify it as an AON (activity-on-node) network, as detailed in the next subsection.

3.1 PERT

As we have seen, the full form of PERT is a program evaluation and assessment technique. With the help of this technology, the project is finished in the least amount of time possible and with minimum resources. In PERT, different activities and events are arranged sequentially and network work is prepared, and the relationships between activities emerge from the network in the head of the project manager. In PERT, a project is divided into different parts, each part is known as an "activity" and the result when each activity is completed is called an "event". By arranging activities and events in a logical order, you get a "network". This network may be called a PERT chart or a PERT chart. Time estimates for various activities are obtained and based on these, the total time required to complete the entire project is determined.

Characteristics of PERT

- A project is to be completed in stipulated time. Hence proper planning of timing of different activities is necessary. In this method the time of the actions may be controlled.
- In this approach, the entire project is segmented into various tasks.
- The tasks are interconnected with each other.
- The anticipated duration of each task is established.
- Each event has preceding activity and succeeding activity except initial event and last event. The initial event has no preceding activity, while the last event has no succeeding activity.
- Once the PERT chart is created, the project's longest path is identified. This is very important in planning.
- In PERT the importance is given to the times of the activities while the costs are ignored.

Advantages of PERT

- PERT is utilized for the planning and management of any project.
- A check can be maintained for completing different activities of the project within the necessary timeframe.
- If the tasks cannot be finished within the allotted time, the float time of the tasks can be utilized.
- The inter relationships of different activities become clear by PERT diagram.
- More attention can be paid to the activities in the path for which the maximum time is used, so the project can be completed in the stipulated time.

- The time can be saved by this method. Over time, the expense can also be lowered.
- All activities must be finished within the specified time, hence proper planning of the manpower can be done.

First of all we have to understand some of the important words used in PERT.

A. Activity

For any project, when time is spent on some work, it is called an activity. In the PERET diagram, each action is depicted by an arrow (→). The direction of the arrow indicates the progress of the operation and the number written on the arrow indicates the time required to complete the operation. If an activity is displayed with the number → 4 it means that 4 units are required for this activity.

B. Event

The result of an action called an event. Each occurrence is depicted by a circle in the PERT diagram. Following the conclusion of the event, other activities begin. How to clean a room is an action, but finishing cleaning is an event. The necessary event or events before the event occurs are called antecedent events. Although an event can also occur after the event has ended, it is called a follow-up event. In a PERT diagram, an occurrence that does not take place is called an initiating event. One or more actions begin from the start event, no activity occurs before this event. Similarly, signs from which no arrow is visible are called the final event. that is, one or more arrows come toward the final event, but no arrows come out. i.e. the operation does not start after the event.

PERT Network

Manage the project effectively; different tools are needed. The internet is among the most crucial and extensively utilized resources.

The diagram obtained by ordering all activities and events is called a network. i.e. network means a schematic representation of activities and events. In the grid, the activities are labeled A, B, C, D, …… etc. and are marked with an arrow (→) and the activity times are shown above the arrows. When an action is completed, an event occurs. The event is depicted by a circle, and the sequence in which the event occurred is shown in the circle. When all activities and events are logically and sequentially connected, they form a network.

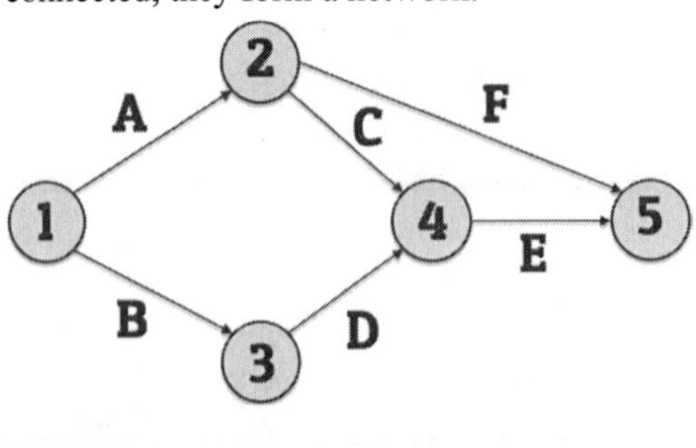

Below are a few key guidelines for constructing a network:

1. One and only one arrow in the network represents each activity.
2. A circle represents an event.
3. An event cannot happen before all activities preceding that event are not completed.
4. An Activity cannot start before the completion of the preceding events.
5. An event cannot happen for two times. i.e. the path of an activity does not make a loop.
6. Each activity starts from an event and ends in an event. The time calculation is done in the direction of arrow from left hand side of the arrow to the right-hand side of the arrow.
7. Each event is given separate number, and these numbers should be given in such way that their sequence is maintained in network diagram.
8. The number of the events should be from left side to the right side in the network diagram.
9. The initial events from which arrows emerge but no arrow comes towards that event is always given 1 number and is shown in the circle as (1)
10. No arrow emerges from the final/last event of the network diagram.
11. It should be noted that the length of the arrow does not indicate the time/ resource duration of that activity. The length and directions of the activity are only symbolic of the time period showing the direction of progress of the activity and the length is so chosen to complete the construction of the network conveniently irrespective of the real time length of the activity.
12. Refrain from using placeholder activity unless necessary to represent the logic.
13. Dotted line arrows represents dummy activities.
14. Prevent looping and intersection of activity arrows by repositioning.
15. Danglers, activities that are isolated must be avoided.
16. For coding purposes, utilize letters for every task including the placeholder task and digits for events.

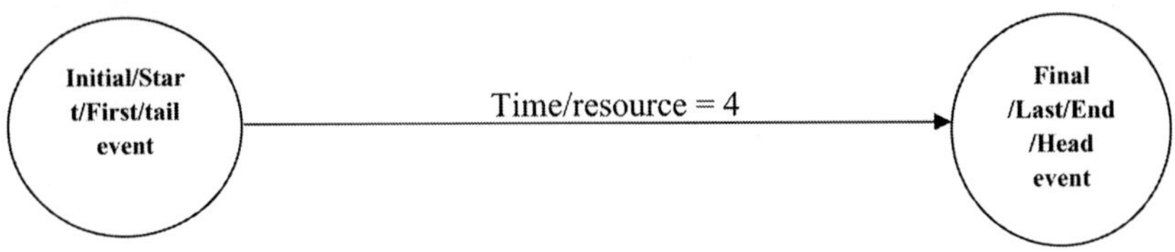

Dummy Activity

An activity which only determines the dependency of one activity over the other, but does not consume any time is called dummy activity. It is represented by dotted line arrow. Dummy activity serves the following purposes:

- To maintain uniqueness in the numbering system, as every activity may have a distinct set of events by which the activity can be identified (a dummy activity has a zero time).
- To maintain the proper logic in the network.

  If two activities have same tail event and same head event with different consumption of time, then in order to maintain the relation we should introduce dummy activity.

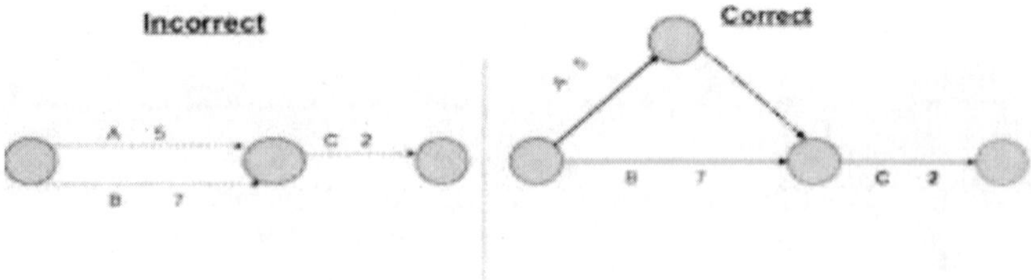

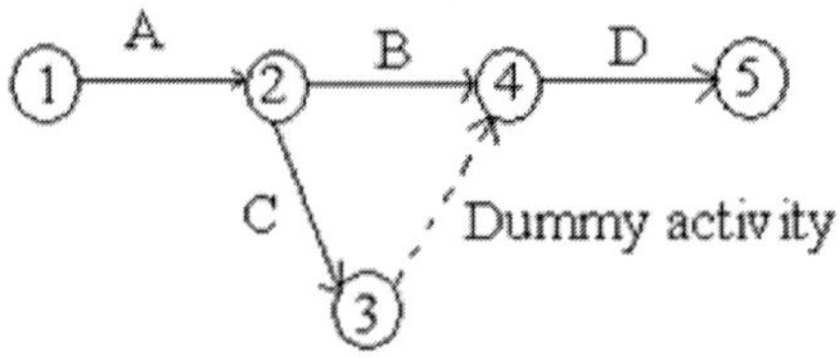

Errors in drawing network diagram

There are three types of errors which are most common in drawing network diagram.

1. **Looping:** Looping error is also referred to as a cycling error in a network diagram. Creating an endless loop in a network is referred to as a looping error. Cycling in a network diagram can occur due to a minor mistake or when, while developing the activity plan, one tries to show the repetition of an activity before beginning the next activity. A closed loop will produce an endless cycle in computer programmes without a built-in-routine for detection or identification of the cycle. Thus, one property of a correctly constructed network diagram is that it is non-cyclic.

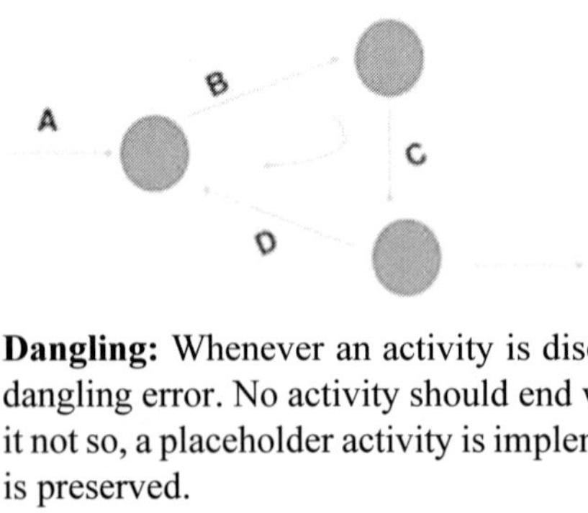

2. **Dangling:** Whenever an activity is disconnected from the network it is called dangling error. No activity should end without being joined to the end event. If it not so, a placeholder activity is implemented to ensure the system's continuity is preserved.

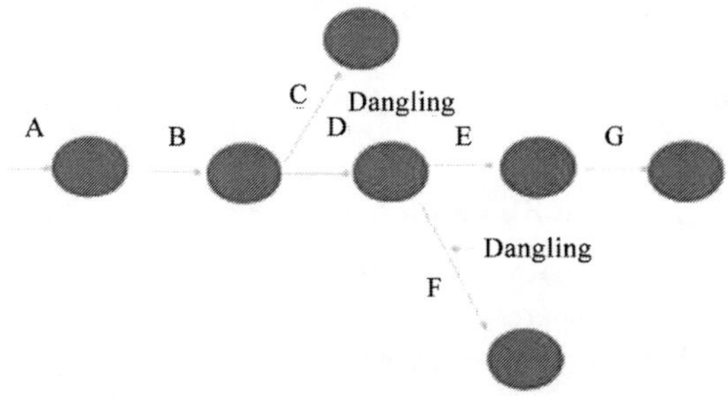

3. **Redundancy:** When the dummy activity is introduced and it is not required then it is called redundancy errors.

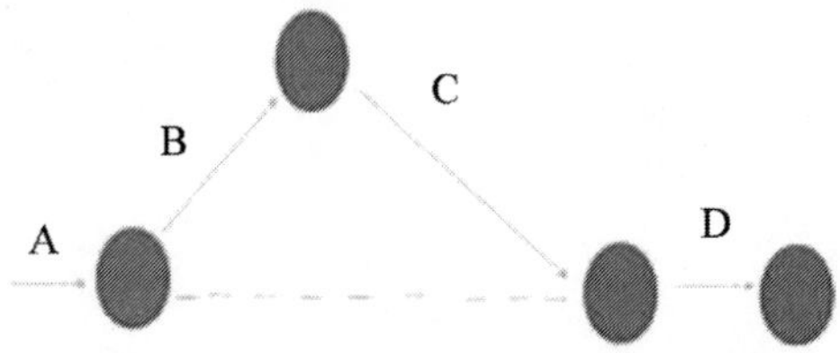

Fulkerson's rule for numbering the events

After the network is drawn in a logical sequence, every c\event is assigned a number which is placed inside the node circle. The number sequence should be such as to reflect the flow of the network. The rule devised by D.R. Fulkerson is used for the purpose of numbering. The steps are the following:

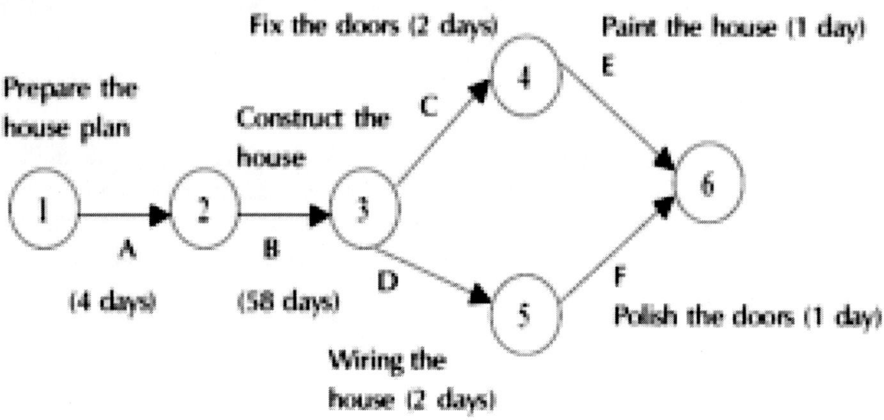

CPM and PERT

- Event identifier must be distinct.
- Event numbering must be done in a sequential manner from left to right.
- The first occurrence that has all outgoing arrows and no incoming arrow is assigned a number 0 to 1.

- The tip of an arrow must consistently display a value greater than the one designated at the base of the arrow.
- Gaps should be left in the order of event numbering to accommodate subsequent inclusion of activities if necessary.

Activity on arrow diagram and activity on node diagram

Both CPM and PERT describe the work plan of project where arrows and circles respectively indicate the actions and occurrences that ought to be completed to reach the project objectives. The activities and events are laid in a planned sequence of their accomplishments. Nevertheless, there are two varieties of notations utilized in the network diagram. They are:

1. **Activity-on-Arrow (AOA)**

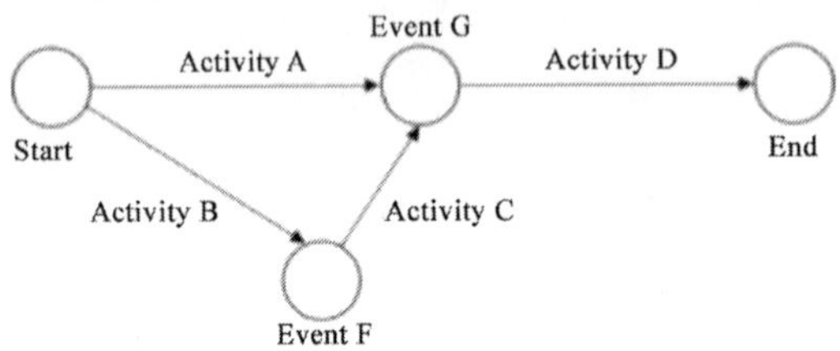

2. **Activity-on-Node (AON)**

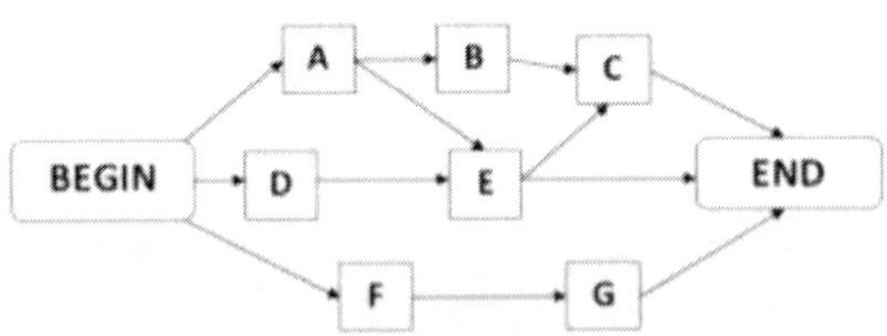

Difference Between PERT and CPM

Project management can be defined as a structured approach to planning, scheduling, executing, monitoring, and controlling various elements of the project to achieve the objectives established during project inception. PERT and CPM represent two network-oriented project management methodologies, which demonstrate the flow and order of tasks and events. Program (Project) Evaluation and Review Technique (PERT) is suitable for projects where the duration required to finish different activities is uncertain. Conversely, Critical Path Method or CPM is suitable for projects that occur repeatedly.

The two scheduling techniques employ a shared strategy for creating the network and determining its critical path. They are utilized for the effective execution of a project and therefore are used together. However, the reality is that CPM differs from PERT in that the former focuses on time, whereas the latter emphasizes time-cost trade-off. Similarly, there are numerous distinctions between PERT and CPM.

Definition of PERT

PERT stands for Program (Project) Evaluation and Review Technique, which involves the planning, scheduling, organizing, coordinating, and controlling of uncertain activities. This technique analyzes and depicts the tasks involved in finishing a project, to determine the least time needed to complete a task and the minimum time necessary to finish the entire project. It was created in the late 1950s. Its purpose is to decrease the time and cost of the project.

PERT employs time as a variable that signifies the scheduled resource utilization along with performance criteria. In this method, the project is initially segmented into tasks and milestones. Subsequently, the correct order is established, and a network is created. Then, the time required for each task is computed, and the critical path (the longest route linking all the milestones) is identified.

Definition of CPM

Created in the late 1950s, the Critical Path Method, or CPM, is a technique employed for the planning, scheduling, coordination, and management of project activities. In this method, it is presumed that the duration of activities is both fixed and definite. CPM is utilized to determine the earliest and latest potential start times for every activity.

The procedure distinguishes between essential and non-essential activities to minimize time and prevent queue formation in the process. The rationale for identifying essential activities is that a delay in any one activity will impact the entire process negatively. This is the reason it is referred to as the Critical Path Method.

In this approach, initially, a list is created that includes all the tasks necessary to finish a project, which is then followed by calculating the time needed for each task to be completed. Subsequently, the relationships among the tasks are assessed. In this context, a 'path' is characterized as a series of tasks within a network. The critical path is identified as the path with the greatest length.

Key Distinctions Between PERT and CPM

The primary distinctions between PERT and CPM are outlined below:

1. PERT is a method for managing projects that involves planning, scheduling, organizing, coordinating, and controlling uncertain activities. CPM is a statistical method of project management in which the planning, scheduling, organizing, coordination, and control of clearly defined activities occurs.
2. PERT is a method for planning and managing time. In contrast to CPM, which is a technique for managing expenses and time.
3. While PERT has developed as a research and development project, CPM has developed as a construction project.
4. PERT is organized based on events, whereas CPM is focused on activities.
5. A deterministic model is utilized in CPM. In contrast, PERT employs a probabilistic model.

6. There are three time estimates in PERT, which are optimistic time (to), most likely time (tm), and pessimistic time (tp). Conversely, there is just one estimate in CPM.
7. PERT method is most effective for highly precise time estimations, whereas CPM is suitable for more general time estimates.
8. PERT addresses uncertain tasks, whereas CPM focuses on predictable tasks.
9. PERT is utilized in situations where the work is non-repetitive. Conversely, CPM pertains to tasks of a repetitive nature.
10. There exists a distinction between critical and non-critical tasks in CPM, whereas this is not true for PERT.
11. PERT is most suitable for research and development initiatives, while CPM is intended for non-research endeavors such as construction projects.
12. Crashing is a compression method utilized in CPM to reduce the length of the project timeline with minimal extra expense. The crashing principle does not apply to PERT.

Table 3.1: Difference between PERT and CPM

Particulrs	**PERT**	**CPM**
Meaning	PERT is a technique for project management that is utilized to oversee uncertain activities within a project.	CPM is a statistical method of project management that oversees clearly defined tasks of a project.
What is it?	A method for organizing and managing time.	A technique to manage expenses and duration.
Orientation	Event-oriented	Activity-oriented
Evolution	Developed as a Research and Development initiative	Evolved as Construction project
Model	Probabilistic Model	Deterministic Model
Focuses on	Time	Time-cost trade-off
Estimates	Three time estimates	One time estimate
Appropriate for	High precision time estimate	Reasonable time estimate
Management of	Unpredictable Activities	Predictable activities
Nature of jobs	Non-repetitive nature	Repetitive nature
Critical and Non-critical activities	No differentiation	Differentiated
Suitable for	Research and Development Project	Non-research endeavors such as civil engineering, boat manufacturing, etc.
Crashing concept	Not Applicable	Applicable

Gantt (Bar) Charts and Microsoft Project® (MSP)

One of the most ancient yet still one of the most effective ways to display project schedule information is the Gantt chart, created around 1917 by Henry L. Gantt, a pioneer in scientific management. The Gantt chart illustrates planned and actual progress for several tasks shown as bars on a horizontal time scale. It serves as a particularly efficient and easily interpretable method of representing the current status for each task in relation to the planned progress for each item in the group. Consequently, the Gantt chart can assist in speeding up, sequencing, and reallocating resources among tasks, in addition to the essential yet ordinary task of monitoring progress. Furthermore, the charts often feature various special symbols to mark or emphasize items of particular significance to the situation being illustrated. Although the 3rd edition of PMBOK now refers to these as "bar charts," we will keep referring to them as Gantt charts in this context.

There are various benefits to utilizing Gantt charts. Firstly, although they can hold a significant amount of information, they are simple to comprehend. While they necessitate regular updates (as does any scheduling/control tool), they are straightforward to manage as long as task requirements remain consistent or substantial changes to the schedule are avoided. Gantt charts offer a visual representation of a project's current situation. Nevertheless, Gantt charts possess a major drawback. If a project is intricate with a large number of activities, it can be very challenging to track numerous activity paths throughout the project. Gantt charts serve as effective instruments for communicating information to senior management, but networks often appear to be more beneficial in the practical management of the project.

Another important aspect of Gantt charts is that they are as simple to create as a network.

As with many subjects, it is crucial for the student to comprehend exactly what networks and Gantt charts represent (and what they do not represent) prior to utilizing MSP or other software to create intricate networks and Gantt charts that the student will need to grasp and utilize. Manually drawing networks and charts is an efficient method to cultivate that comprehension. However, once understanding is achieved, software becomes simpler, quicker, and when applied to a project of a scale that mirrors reality, significantly more economical.

Consider the illustration in Figure 3.1 for MSP. If we launch MSP, we encounter the format that is used for inputting action plan (Work Breakdown Structure (WBS)) information into the application. (You might need to close the MSP help window to view the complete form). Inputting the information is simple. We start by entering an activity called "START. " We designate it a duration of 0 days, which categorizes it as a "milestone" instead of a genuine "activity. " We then input activity a with a duration of five days and proceed with the remaining activities. At the end of the list, we include FINISH with a duration of zero days, marking the project's concluding milestone.

Tasks	*Precedence*	*Time*	*Cost*	*Who Does*
a	—	5 days	—	—
b	—	4 days	—	—
c	a	6 days	—	—
d	b	2 days	—	—
e	b	5 days	—	—
f	c,d	8 days	—	—

Figure 3.1 Sample WBS (Work Breakdown Structure)

The software automatically generates a WBS (ID) number for each activity upon entry. You have the option to delete or add columns if desired. If you do not specify a start date, the MSP will automatically set the current date as the start date. While inputting data from the action plan, MSP will automatically create an AON network and a Gantt chart. (The Gantt chart will appear on the right side of the action plan details). If the names and durations of activities are entered without including the relevant predecessor information, all activities will be presumed to commence on the same start date. As you input the predecessor information, the correct relationships between the activities will be displayed; refer to Figures 3.2 and 3.3.

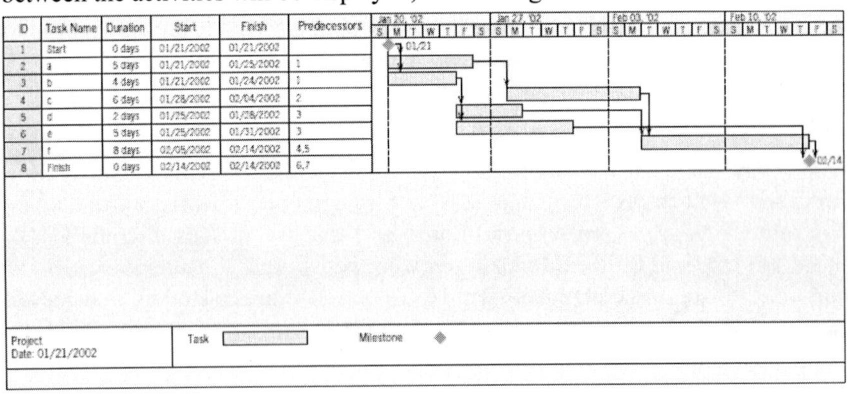

Figure 3.2: MSP schedule and Gantt chart for example project in Figure 3.1

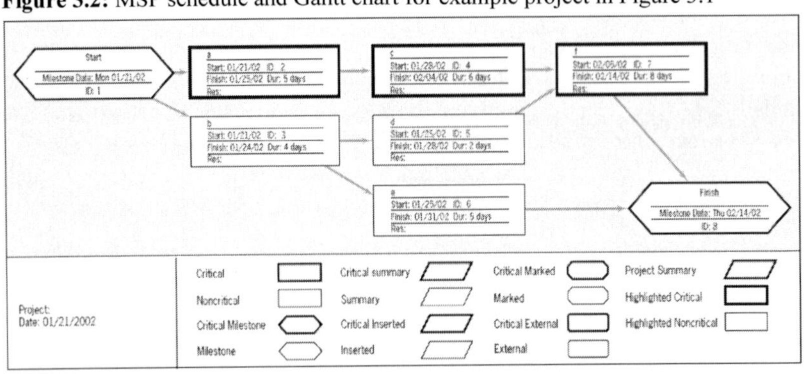

Figure 3.3: MSP AON (Activity On Node) network for the example project in Figure 3.1.

Our emphasis up to this point has merely been to illustrate the technological dependencies found within a network or Gantt chart. A brief examination of the AON network or the Gantt chart uncovers something significant. If we tally the duration of activities for all tasks in the action plan, we find that there are 30 days of work to be scheduled. Nonetheless, as indicated by the network and chart, the project is scheduled to commence on January 21, 2002, and will wrap up on February 14, 2002. This results in 25 days, not 30. Furthermore, MSP defaults to a work calendar featuring a five-day workweek. (Note that Saturday and Sunday are lightly shaded on the Gantt chart to indicate these are nonworking days). Should the calendar be adjusted to a seven-day week, the project will only require 19 days to be completed. It could potentially be finalized by February 8. The assessment of project duration has not yet been considered, yet it will be addressed very shortly. Manual calculations are not exceedingly challenging, but MSP can execute them more swiftly and effortlessly. It is important, however, to remember that software makes assumptions regarding factors like the number of days worked in a week, the hours worked each day, and various other details that will be explored later. These assumptions can be modified if desired, but they must be considered each time the software is used to outline a project.

This instance illustrates both the benefits and drawbacks of the Gantt chart. Its main benefit is that it is easy to comprehend. All commonly utilized project management software is capable of generating Gantt charts, and many provide various customization options. Overall, the simplicity of creation and usage has resulted in the Gantt chart being the preferred approach for presenting a project timeline. However, an AON network continues to be the most efficient method for the project manager to keep track of the timeline, and the reader could be misled if the Gantt chart is not scrutinized carefully or if it is missing any crucial details (Wilkens, 1997).

Slack (aka, Float)

We will now focus on the latest starting times (LS) for the activities. As stated in the previous section, the ES for an activity is equal to the maximum EF for its preceding activities. A critical question for the PM is this: What is the latest time (LS) that activity i can start without causing a delay in the entire project?

Refer to Figure 3.4. The project spans a crucial period of 43 days. Activity i must consequently be finished by day 43, as indicated by LF (latest finish time), found at the bottom right of its node. Additionally, activity i takes 18 days to finish. Therefore, I must be started no later than day 25 (43 – 18 = 25) in order for the project to conclude on day 43. The LS for activity i is therefore 25, indicated at the bottom left corner of the node. Since i cannot commence until activities g and h are finished, the latest finish time (LF) for both is still day 25. The difference between the LS and the ES for any activity is called its float or slack. In the case of activity i, it must begin no later than day 25 but could start as early as day 24, leading to one day of slack. It should be clear that all activities along the critical path have zero slack. * Activities that have zero slack cannot be delayed without making the project late.

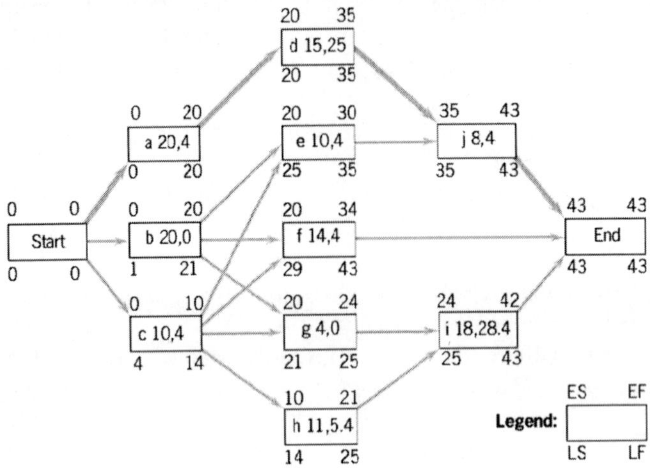

Figure 3.4: AON (Activity On Node)network showing earliest and latest start and finish time and critical path(arrow in bold)

For another example, consider activity f. Its ES is on day 20, which aligns with the EF of its prior activity b. The LS for activity f is calculated as 43 – 14 = 29. If f begins after day 29, it will delay the whole project. Activity f has slack of LS – ES = 29 – 20 = 9 days.

To determine the slack for any task, we conduct a reverse pass (from right to left) through the network, similar to how we performed a forward pass (left to right) to identify the critical path and duration, along with the ESs and EFs for subsequent activities. There is one straightforward guideline that we need to follow: When encountering two or more nonessential tasks along a route, it is standard practice to compute the cushion for every assignment as if it were the sole task on that path. Therefore, when calculating the padding for each task as if it were the i's predecessors are experiencing delays. Naturally, if some task, x had six days of flexibility, and if a prior task was delayed, causing the event to be pushed back by two days, then task x would only have four days of flexibility remaining, having lost two days because of the previous delay.

It is simple to calculate slack for tasks that are immediate predecessors of the final node. As we move to earlier tasks, it becomes somewhat more intricate. Consider activity g. Remembering our assumption that the other tasks in the same path do not use any of the available slack, we see that activity i must follow g, and that g takes place after activities b and c. Starting with the LS of activity i at 25, we subtract four days for g (25 - 4 = 21). Thus, g cannot begin later than day 21 without delaying the network. The ES for g is day 20, indicating that g has one day of slack.

As a different illustration, take a look at activity e. Activity e is required to be finished by day 35, which is the LS of activity j. Consequently, the LS for e is calculated as 35 - 10 = 25. The ES for task e is day 20, allowing it to have five days of slack. Table 3.1 presents the LS, ES, and slack for every activity.

Table 3.1. Durations and Slacks for Network in Figure 3.4

Activity	*LS*	*ES*	*Slack*
a	0	0	0
b	1	0	1
c	4	0	4
d	20	20	0
e	25	20	5
f	29	20	9
g	21	20	1
h	14	10	4
i	25	24	1
j	35	35	0

On some occasions, the PM might agree on a suitable completion date for a task that provides some leeway within the whole network. For example, if a feasible date was 50 business days following the project's initiation, then the network would possess an overall total of 50 - 43 = 7 days of leeway.

Some authors and MSP distinguish between "overall" slack or float and "available" slack or float. Total slack is calculated as LF – EF or LS – ES as outlined previously. Free slack is characterized as the duration an action may be postponed without influencing the start time of any subsequent activity. Activity h may be delayed three days without impacting the beginning time of activity i. Activity h has three days of available slack and four days of overall slack.

Precedence Diagramming

One limitation of the AOA (Activity On Arrow) network technique is that it does not permit for leads and lags between two tasks without significantly raising the number of sub activities needed to accommodate this. In construction projects specifically, it is very typical for the following constraints to arise. (Node labels are illustrated in Figure 3.5)

Activity 2 in a Finish to Start arrangement must not commence until Activity 1 has been finished. This is the standard configuration of an activity along with its former version. Other arrangements of finish-start sequences are also feasible. If the former version details had been stated as "1FS + 2 days," Activity 2 would be planned to initiate at a minimum two days after the conclusion of Activity 1, as illustrated in Figure 3.5. For example, if Activity 1 involved the casting of a concrete pathway, Activity 2 could be any task that utilized the walkway.

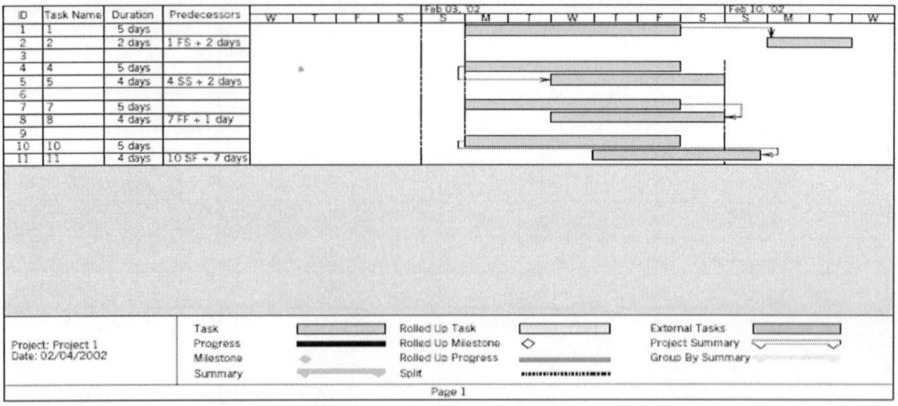

Figure 3.5: Precedence diagramming conventions.

Start to Start Activity 5 is unable to commence until Activity 4 has been in progress for a minimum of two days. The positioning of electrical wires cannot initiate until two days following the commencement of framing.

Finish to Finish Activity 7 must be completed at minimum one day before Activity 8 is finished. If Activity 7 involves preparing the walls of a house, then Activity 8 could involve the tasks related to choosing, buying, and ultimately delivering the wallpaper. It is crucial not to install the wallpaper until the wall primer has dried for 24 hours.

Activity 11 cannot be completed until a minimum of 7 days have elapsed since the start of Activity 10. If Activities 10 and 11 constitute the two main cruising activities during a prepaid weeklong ocean cruise, the overall duration must be no less than the guaranteed week. The S–F connection is rare because of the generally simpler ways to establish the required relationship.

Precedence diagramming is a technique using AON networks that easily integrates these leads and lags within the network. MSP handles leads and lags with ease. The times for network nodes are calculated in a manner similar to Activity On Node (AON) / (AOA) Activity on Arrow times. Because of the constraints of leads and lags, it is frequently advantageous to develop a Gantt chart to observe what is genuinely happening. The broader array of precedence relationships that AON facilitates is pertinent for a variety of projects, particularly those in construction. [For further details on this approach, consult Al-Hammed *et al.* (1998).] Most contemporary project management software will accommodate leads, lags, delays, and various other constraints in their standard AON network and Gantt chart features.

Example 1

Create a network diagram for the project outlined below:

Activity	Immediate Predecessor Activity
A	-
B	A

C, D	B
E	C
F	D
G	E, F

Solution

Activity A does not have any preceding activities. In other words, it is the initial activity. Assume that activity A moves the project from event 1 to event 2. Consequently, we have the following depiction for A:

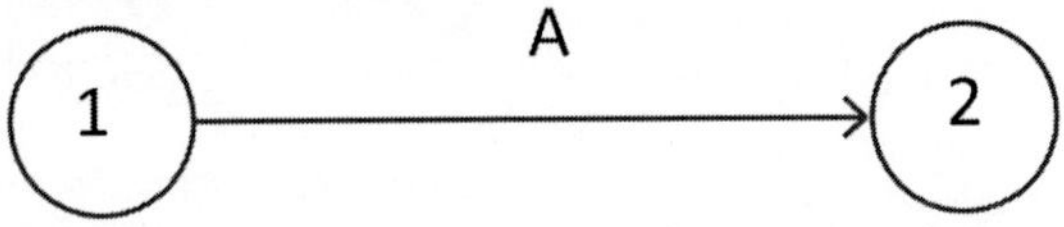

For activity B, activity A is the predecessor. Let us assume that B connects nodes 2 and 3. Consequently, we obtain.

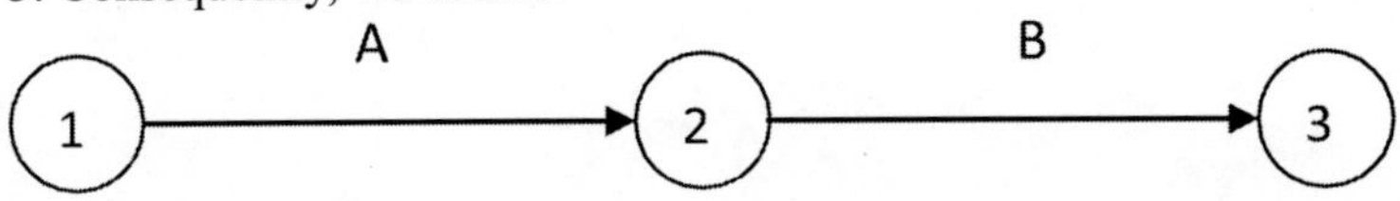

Activities C and D have B as the prior activity. Consequently, we derive the following:

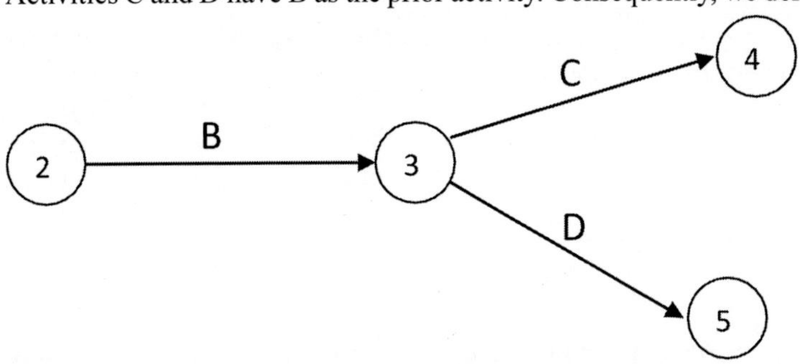

Activity E has C as the preceding activity.. Therefore, we obtain

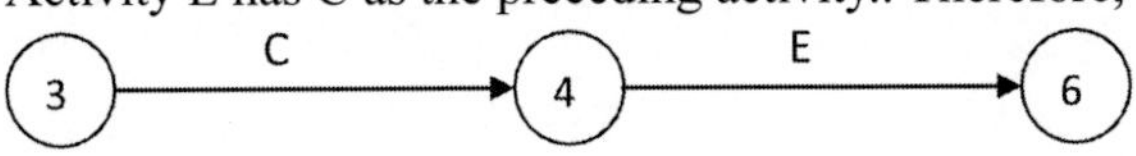

Activity F has D as the preceding activity. Thus we obtain

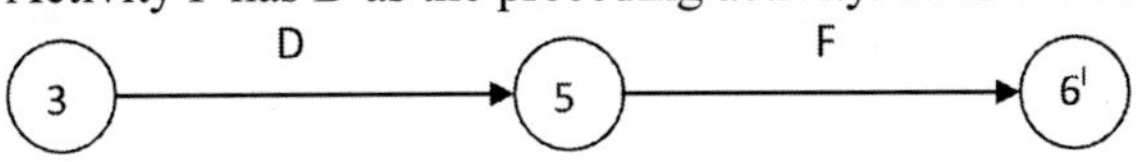

Activity G has E and F as preceding activities. This can occur only if nodes 6 and 6' are identical. Therefore, rename node 6' to node 6. After this, we obtain

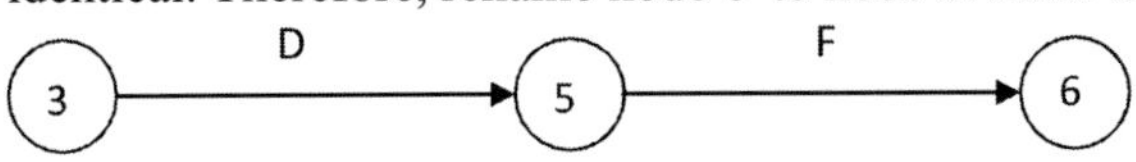

and

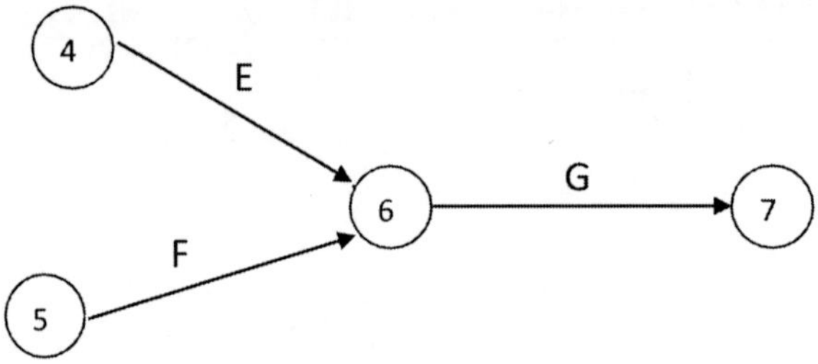

G is the last activity.

Combining all the components, we arrive at the subsequent diagram of the project network:

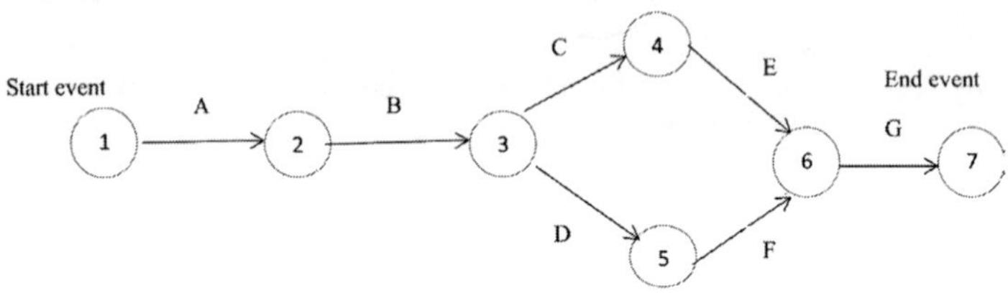

The diagram is confirmed by consulting the provided data.

Note: A significant aspect can be noted for the diagram above. Take into account the subsequent components in the diagram.

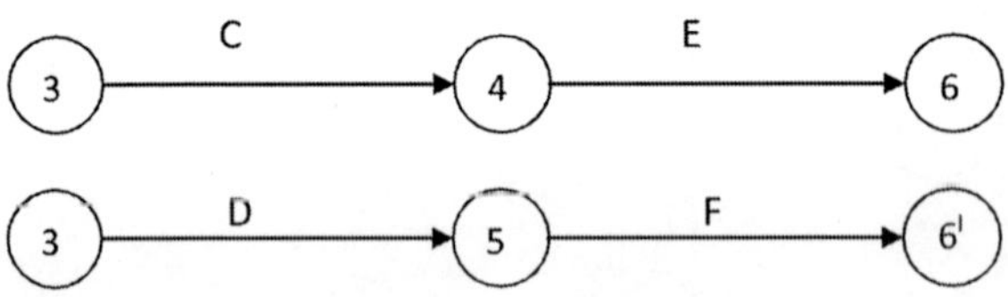

and

We considered nodes 6 and 6' to be identical. Instead, we can keep them as separate nodes. Next, to ensure connectivity in the network, we connect nodes 6' and 6 through a dummy activity. This leads us to the following diagram for the project network:

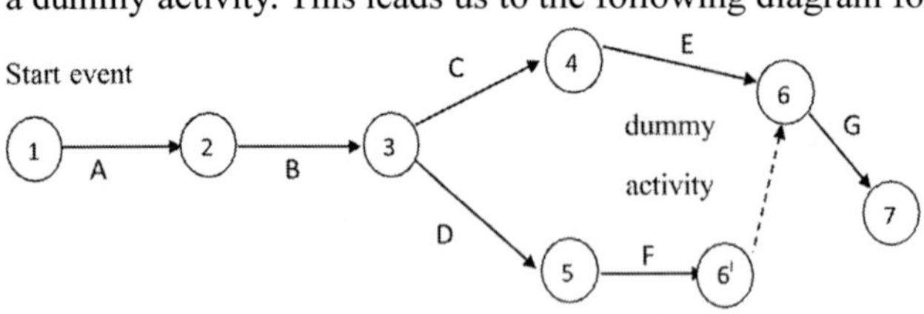

Build the network diagram for the subsequent project:

Activity	Immediate Predecessor Activity
A	-
B	-

C	A
D	B
E	A
F	C, D
G	E
H	E
I	F, G
J	H, I

Critical Path Method (CPM)

Lesson Outline

The ideas of critical path and critical activities. Position of the critical path. Assessment of the project completion duration

Learning Objectives

After completing this lesson you ought to be capable of comprehend the meanings of critical path and critical activities recognize critical path and critical activities establish the project completion time

Introduction

The critical path method (CPM) focuses on identifying the duration required to finish a project and the key activities that a manager should concentrate on.

Assumption for CPM

In CPM, it is presumed that an exact time estimation is accessible for every activity.

Project Completion Time

From the initial event to the final event, the duration needed to finish all the tasks of the project in the designated order is referred to as the project completion duration.

Path in A Project

A continuous series, made up of nodes and activities in turn, starting with the beginning event and concluding at the finishing event of a network is referred to as a path in the network.

Critical Path and Critical Activities

Consider every route in a project, starting at the initial event and concluding at the final event. For each route, determine the duration of execution by summing the time for the separate activities along that route.

The route with the longest duration is referred to as the critical path, and the tasks along this route are known as critical activities or bottleneck activities. These activities are termed critical because they cannot be postponed. On the other hand, a non-critical activity may be postponed to a certain degree. Any postponement in a critical activity will result in delaying the entire project. In contrast, a permissible delay in a non-critical activity will not affect the project's overall completion. It is important to note that a delay in a non-critical activity that exceeds a threshold would indeed delay the entire project. Occasionally, there may be multiple critical paths for a single project. A project manager must give particular attention to critical activities.

Example 2

The subsequent information is accessible concerning a project:

Activity	Predecessor Activity	Duration (Weeks)
A	-	3
B	A	5
C	A	7
D	B	10
E	C	5
F	D,E	4

Establish the critical path, identify the critical activities, and calculate the project completion time.

Solution

First, let's create the network diagram for the specified project. We denote the time estimates on the arrows that signify the activities. We arrive at the following diagram:

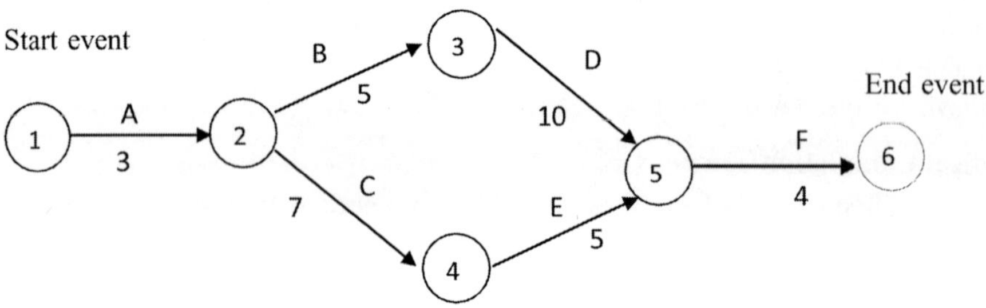

Consider the routes, commencing with the start node and ceasing with the end node. There exist two such routes for the specified project. They are as follows:

Path I

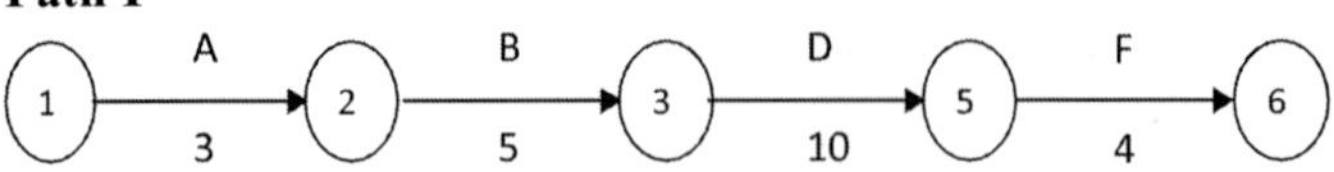

with a duration of 3 + 5 + 10 + 4 = 22 weeks.

Path II

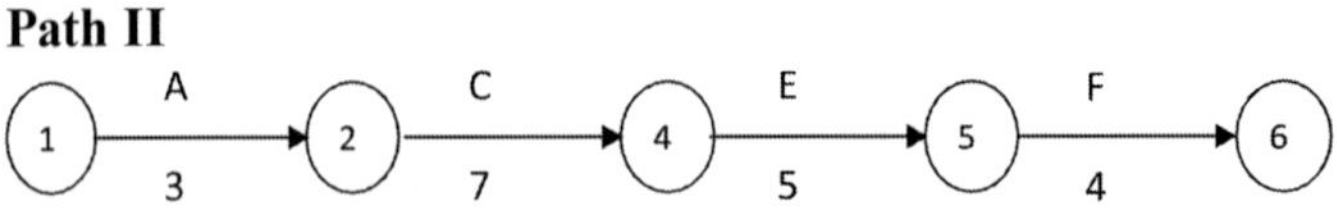

with a duration of 3 + 7 + 5 + 4 = 19 weeks.

Compare the durations for the two routes. Maximum of {22,19} = 22.

We observe that path I has the highest duration of 22 weeks. Consequently, path I is identified as the critical path.

The essential tasks are A, B, D, and F. The duration for project completion is 22 weeks.

We observe that C and E are non-essential activities.

Time for path I - Time for path II = 22 - 19 = 3 weeks.

Consequently, collectively the non-critical activities may be postponed for up to a maximum of 3 weeks, without postponing the overall completion of the entire project.

Example 3

Determine the completion time and the key activities for the subsequent project:

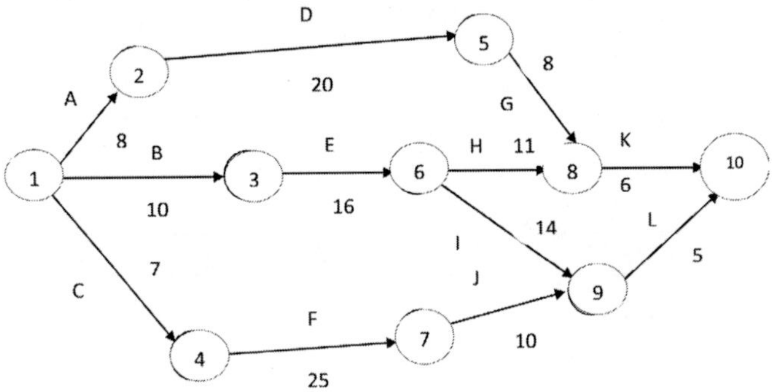

Solution

In total, we recognize 4 routes, commencing from the initial node of 1 and concluding at the final node of 10. They include the following:

Path I

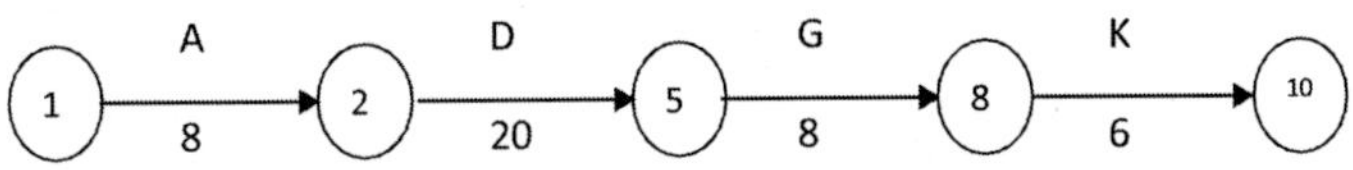

Time for the route = 8 + 20 + 8 + 6 = 42 time units.

Path II

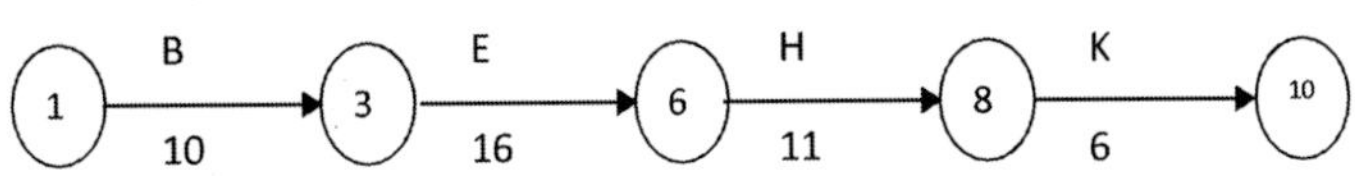

Time for the route = 10 + 16 + 11 + 6 = 43 units of time.

Path III

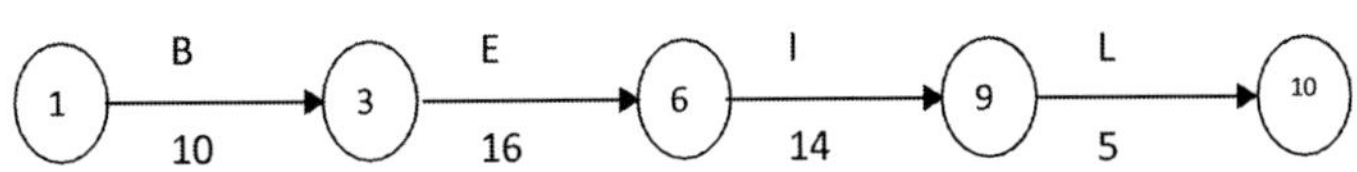

Time for the route = 10 + 16 + 14 + 5 = 45 units of time.

Path IV

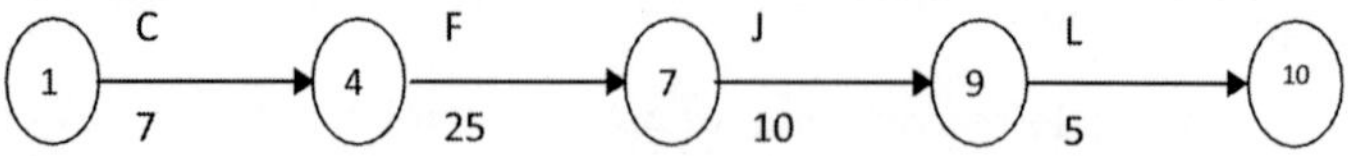

Time for the route = 7 + 25 + 10 + 5 = 47 units of time.

Compare the durations for the four routes. The maximum of {42, 43, 45, 47} is 47. It is evident that the path listed below has the longest duration, making it the critical path:

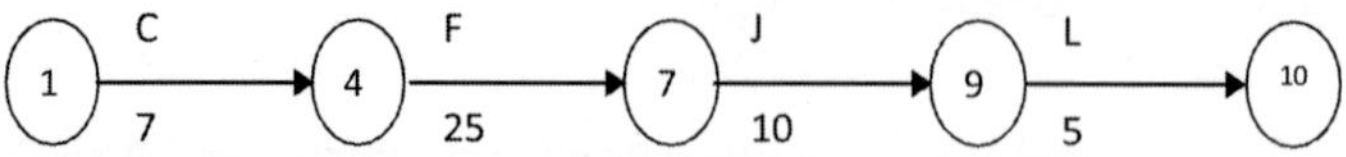

The essential tasks are C, F, J, and L. The peripheral tasks are A, B, D, E, G, H, I, and K. The duration for completing the project is 47 time units.

Example 4

Draw the network diagram and identify the critical path for the subsequent project:

Activity	Time estimate (Weeks)
1- 2	5
1- 3	6
1- 4	3
2 -5	5
3 -6	7
3 -7	10
4 -7	4
5 -8	2
6 -8	5
7 -9	6
8 -9	4

Solution

We possess the subsequent network diagram for the project:

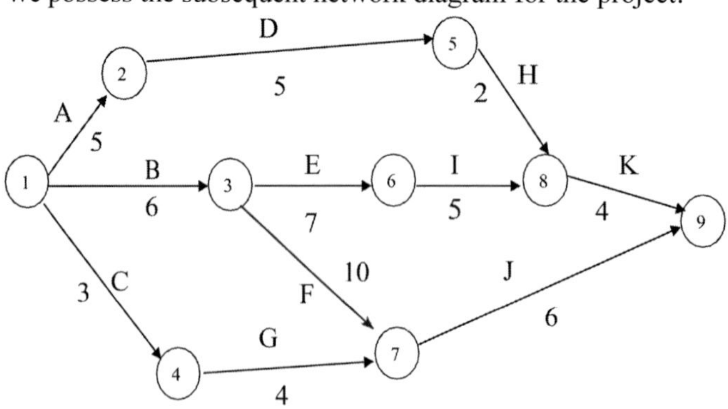

Solution

We claim that there exist 4 routes, commencing at the starting node of 1 and concluding at the ending node of 9. They are listed as follows:

Path I

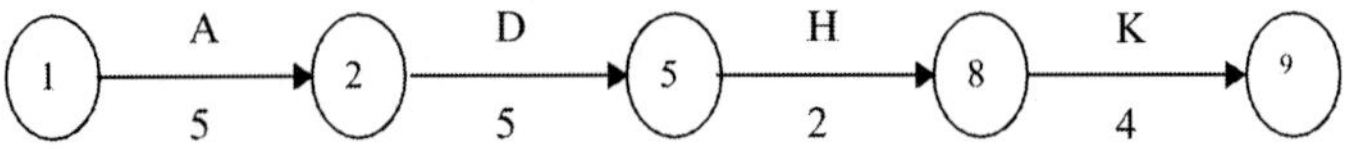

Time for the route = 5 + 5 + 2 + 4 = 16 weeks.

Path II

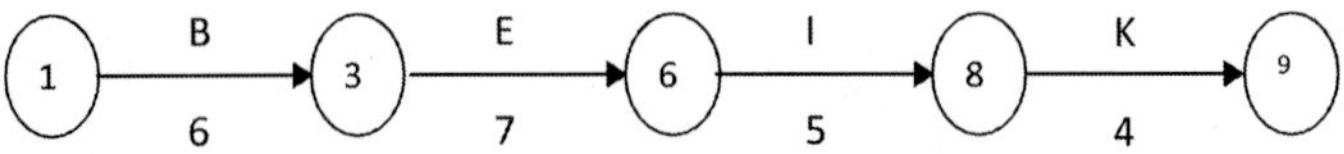

Time for the route = 6 + 7 + 5 + 4 = 22 weeks.

Path III

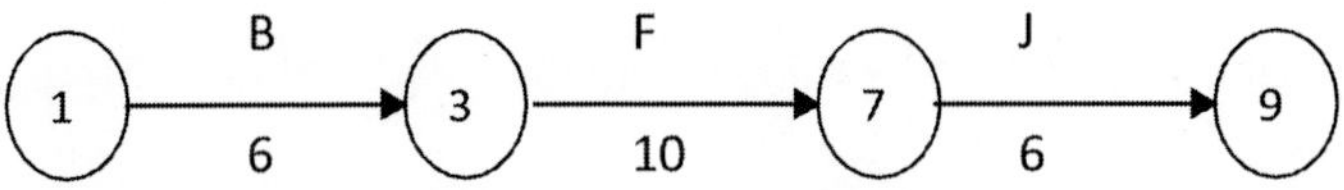

Time for the route = 6 + 10 + 6 = 22 weeks.

Path IV

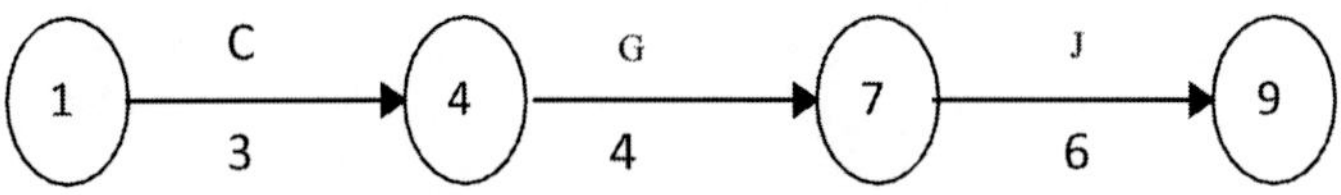

Time for the route = 3 + 4 + 6 = 13 weeks.

Compare the durations for the four routes. Maximum of {16, 22, 16, 13} = 22. We observe that the subsequent route has the highest duration, thus it is the critical path:

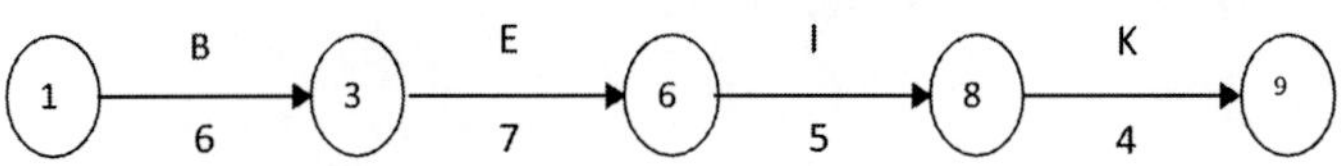

The important activities are B, E, I and K. The unimportant activities are A, C, D, F, G, H and J. The duration for completing the project is 22 weeks.

Below are the time estimates along with the precedence relationships of the tasks in a project network:

Activity	Immediate Predecessor Activity	Time estimate (weeks)
A	-	4
B	-	7
C	-	3
D	A	6
E	B	4

F	B	7
G	C	6
H	E	10
I	D	3
J	F, G	4
K	H, I	2

Draw the diagram of the project network. Identify the critical path and the duration for project completion.

PERT

Lesson Outline

The Idea of Pert

Assessments of the Duration of an Activity

Identification of Critical Path

Learning Objectives

After completing this lesson, you ought to have the ability to

1. Comprehend the significance of PERT
2. Identify the critical path
3. Ascertain the project completion time
4. Determine the likelihood of finishing a project ahead of a specified time
5. Utilize the normal probability distribution table

Introduction

Programme Evaluation and Review Technique (PERT) is an instrument that would assist a project manager in planning and controlling a project. It would facilitate him in consistently overseeing a project and implementing corrective actions whenever required. This method includes statistical techniques.

Assumptions for PERT

Note that in CPM, it is assumed that an accurate time estimate is available for every activity in a project. However, it is often found that this is not feasible in practice.

In PERT, it is assumed that obtaining exact time estimates for every activity is not feasible, and instead, only probabilistic time estimates can be achieved. Here, a multiple time estimate method is adopted. In the context of probabilistic time estimates, the following three types of estimates can be made:

Pessimistic time assessment (tp)

Optimistic time assessment (to)

Most probable time assessment (tm)

The hopeful time estimate relies on the belief that an activity will be free of difficulties while being completed quickly. Conversely, a negative time estimate is based on the expectation that unforeseen issues will arise during the execution of an activity, leading to increased time consumption. The most probable time estimate falls between the optimistic and pessimistic time estimates. Therefore, the three time estimates are interrelated.

$to \leq tm \leq tp$

Practically speaking, neither the negative nor the positive estimate may be accurate in reality, and it is the most probable time estimate that is anticipated to dominate in nearly all situations. Consequently, it is better to prioritize the most probable time estimate more heavily.

We assign a weight of 4 to the most likely time estimate and a weight of 1 to both the pessimistic and optimistic time estimates. We calculate a time estimate (te) as the weighted average of these estimates in this manner:

$$t_e = \frac{t_o + 4t_m + t_p}{6}$$

Since we have utilized 6 units (1 for tp, 4 for tm, and 1 for to), we calculate the total by dividing the sum by 6. Using this time estimate, we can ascertain the project completion duration as relevant for CPM.

Since PERT consists of averaging three time estimates for every activity, this approach is quite practical, and the outcomes from PERT will possess a fair degree of reliability.

Measure of Certainty

The three assessments of time are arranged in a manner that

$to \leq tm \leq tp$

Consequently, the span for the time assessment is $t_p - t_o$.

The duration of an activity in a project network adheres to a distribution with a standard deviation roughly equal to one sixth of the range.

i.e., The standard deviation $= \sigma = \frac{t_p - t_o}{6}$

and the variance $= \sigma^2 = \left(\frac{t_p - t_o}{6}\right)^2$

The reliability of the time prediction for an activity can be evaluated using the variance. The higher the variance, the more unpredictability there is in the time prediction of an activity.

Example 5

Two specialists A and B assessed an activity and reached the subsequent time evaluations.

Expert	Time Estimate		
	t_o	t_m	t_p
A	4	6	8
B	4	7	10

Ascertain which specialist exhibits greater confidence regarding his time estimations:

Solution

Variance (σ^2) in time estimates $= \left(\frac{t_p - t_o}{6}\right)^2$

In the case of expert A, the variance $= \left(\frac{8-4}{6}\right)^2 = \frac{4}{9}$

As regards expert B, the variance $= \left(\frac{10-4}{6}\right)^2 = 1$

Thus, the variance is smaller in the situation of A. Therefore, it can be concluded that the expert A is more confident about his time estimates.

Determining the Completion Time of a Project in PERT

Example 6

Determine the duration necessary to finish the subsequent project and the essential tasks:

Activity	Predecessor Activity	Optimistic time estimate (to days)	Most likely time estimate (tm days)	Pessimistic time estimate (tp days)
A	-	2	4	6
B	A	3	6	9
C	A	8	10	12
D	B	9	12	15
E	C	8	9	10
F	D, E	16	21	26
G	D, E	19	22	25
H	F	2	5	8
I	G	1	3	5

Solution

From the three-time estimates tp, tm, and to, determine te for each activity. We get the subsequent table:

Activity	Optimistic time estimate (to)	4 x Most likely time estimate	Pessimistic time estimate (tp)	to+ 4tm+tp	Time estimate $te = \frac{t_o + 4t_m + t_p}{6}$
A	2	16	6	24	4
B	3	24	9	36	6
C	8	40	12	60	10
D	9	48	15	72	12
E	8	36	10	54	9
F	16	84	26	126	21
G	19	88	25	132	22
H	2	20	8	30	5
I	1	12	5	18	3

By utilizing the individual time estimates for the activities, we obtain the subsequent network diagram for the project.

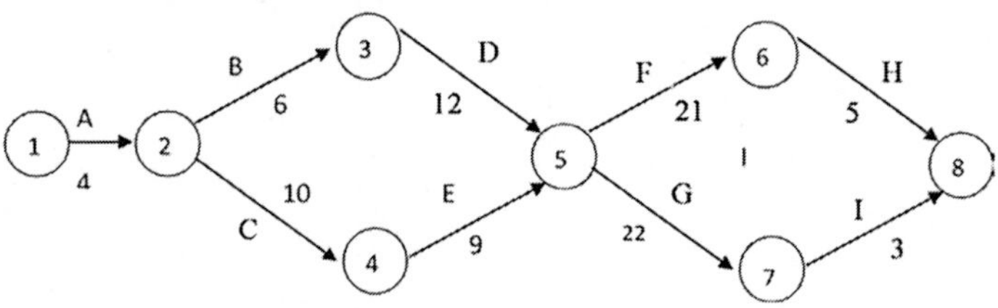

Consider the routes, starting with the initial node and concluding with the final node. There are four such routes for the specified project. They are as follows:

Path I

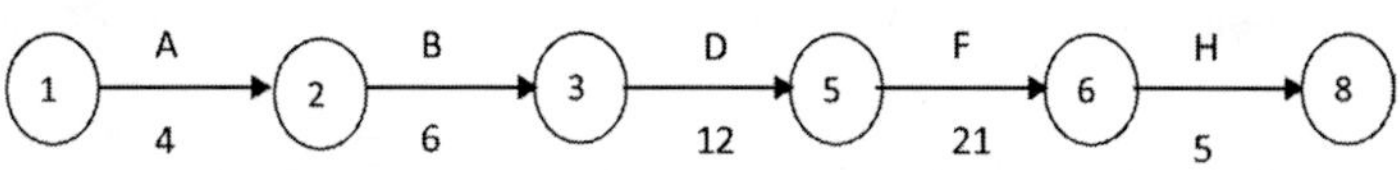

Time for the route: 4+6+12+21+5 = 48 days.

Path II

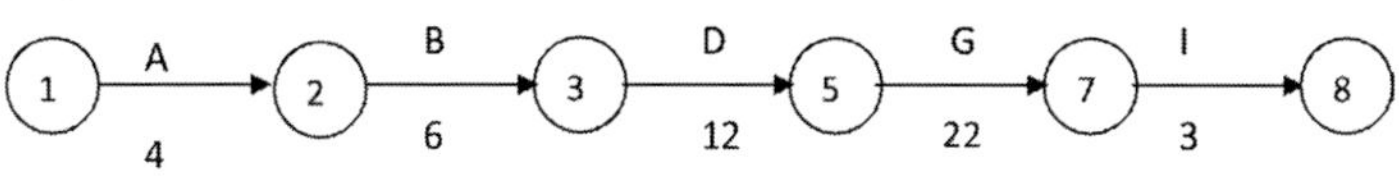

Time for the route: 4+6+12+22+3 = 47 days.

Path III

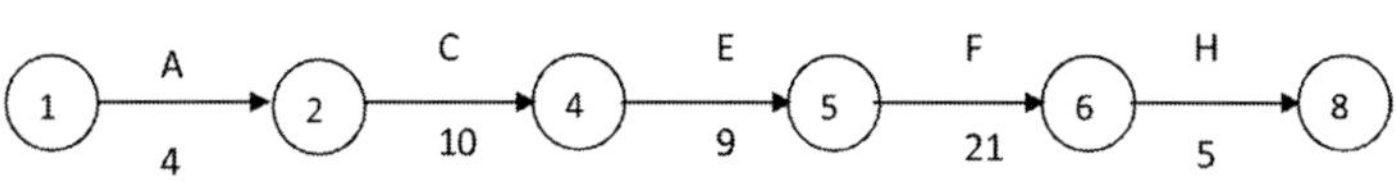

Time for the route: 4+10+9+ 21+5 = 49 days.

Path IV

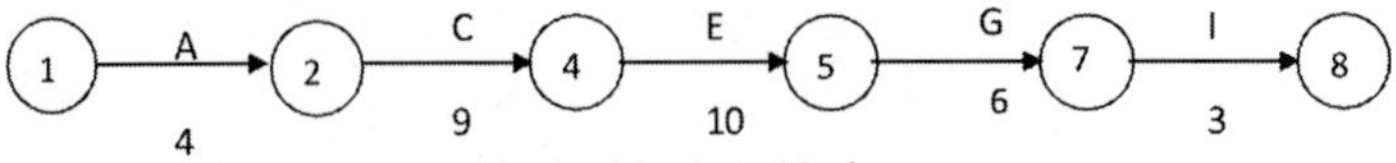

Time for the route: 4+10+9+22+3 = 48 days.

Compare the durations for the four routes.

Maximum of {48, 47, 49, 48} = 49.

We observe that Path III has the highest duration.

Thus, the critical path is Path III. i.e., 1 2 4 5 6 8.

The critical tasks are A, C, E, F and H.

The non-critical tasks are B, D, G and I.

Project duration (Also known as project length) = 49 days.

Example 7

Determine the time, variance, and standard deviation of the project using the following time estimates in weeks:

Activity	Optimistic time estimate (to)	Most likely time estimate (tm)	Pessimistic time estimate (tp)
1-2	3	6	9
1-6	2	5	8
2-3	6	12	18
2-4	4	5	6
3-5	8	11	14
4-5	3	7	11
6-7	3	9	15
5-8	2	4	6
7-8	8	16	18

Solution

From the three time estimates t p, tm and to , determine te for every activity. We obtain the subsequent table:

Activity	Optimistic time stimate (to)	4 x Most likely time estimate	Pessimistic time estimate (tp)	to + 4tm + tp	Time estimate $t_e = \frac{t_o + 4t_m + t_p}{6}$
1-2	3	24	9	36	6
1-6	2	20	8	30	5
2-3	6	48	18	72	12
2-4	4	20	6	30	5
3-5	8	44	14	66	11
4-5	3	28	11	42	7
6-7	3	36	15	54	9
5-8	2	16	6	24	4
7-8	8	64	18	90	15

With the individual time projections of the tasks, we obtain the subsequent network diagram for the project.

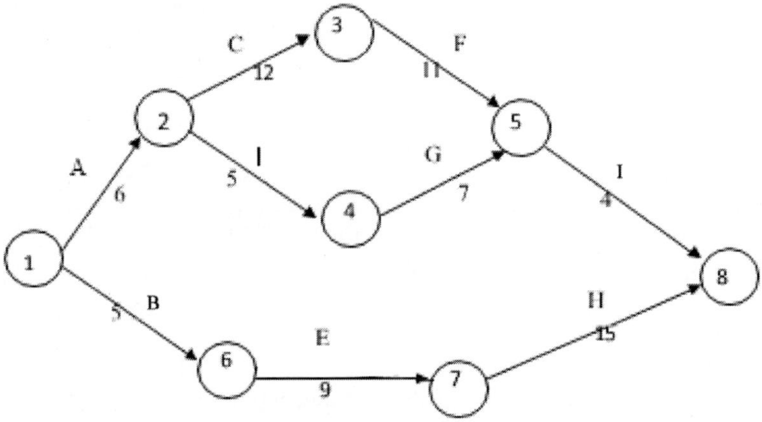

Consider the routes, starting with the initial node and concluding with the final node. There are three such routes for the specified project. They are as follows:

Path I

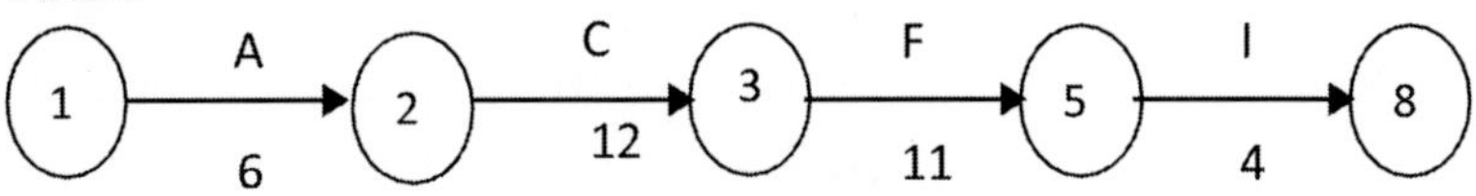

Time for the route: 6+12+11+4 = 33 weeks.

Path II

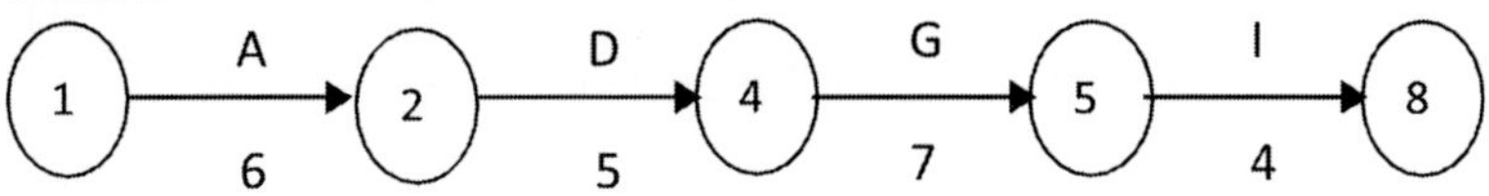

Time for the journey: 6+5+7+ 4= 22 weeks.

Path III

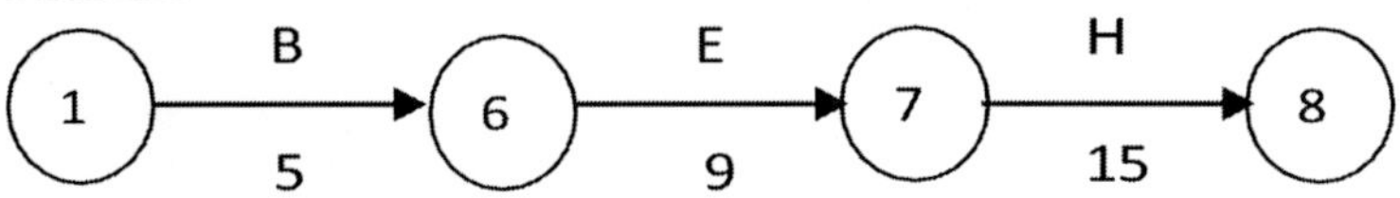

Time for the route: 5+9+15 = 29 weeks.

Compare the durations for the three routes.

Maximum of {33, 22, 29} = 33.

It is observed that Path I has the longest duration.

Hence, the critical path is Path I. i. e. , 1 2 3 5 8

The critical tasks are A, C, F, and I.

The activities that are not critical are B, D, G, and H.

Total project duration = 33 weeks.

Calculation of Variance and Standard Deviation for the Essential Activities:

Critical Activity	Optimistic time estimate (to)	Most likely time estimate (tm)	Pessimistic time estimate (tp)	Range (tp - to)	Standard deviation $\sigma = \frac{t_p - t_o}{6}$	Variance $\sigma^2 = \left(\frac{t_p - t_o}{6}\right)^2$
A: 1 2	3	6	9	6	1	1
C: 2 3	6	12	18	12	2	4
F: 3 5	8	11	14	6	1	1
I: 5 8	2	4	6	4	2/3	4/9

Variance of project duration (Also referred to as Variance of project length) =

Total of the variances for the critical activities = 1+4+1+ 4/9 = 58/9 Weeks.

Standard deviation of project duration = √Variance = √58/9 = 2. 54 weeks.

Example 8

A project is made up of seven activities with the subsequent time estimates. Determine the likelihood that the project will finish in 30 weeks or fewer.

Activity	Predecessor Activity	Optimistic time estimate (to days)	Most likely time estimate (tm days)	Pessimistic time estimate (tp days)
A	-	2	5	8
B	A	2	3	4
C	A	6	8	10
D	A	2	4	6
E	B	2	6	10
F	C	6	7	8
G	D, E, F	6	8	10

Solution

From the three time predictions, and, compute for every task. The outcomes are provided in the table below:

Activity	Optimistic time estimate (to)	4 x Most likely time estimate	Pessimistic time estimate (tp)	to + 4tm+tp	Time estimate $t_e = \frac{t_o + 4t_m + t_p}{6}$
A	2	20	8	30	5
B	2	12	4	18	3
C	6	32	10	48	8
D	2	16	6	24	4
E	2	24	10	36	6
F	6	28	8	42	7
G	6	32	10	48	8

With the individual time projections of the tasks, the ensuing network diagram is created for the project.

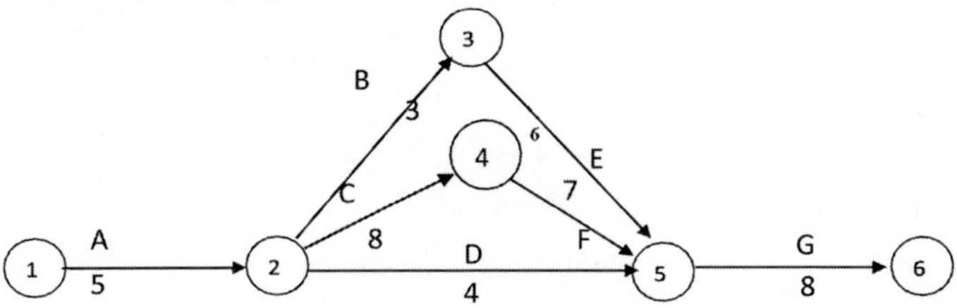

Consider the routes, starting from the initial node and concluding at the final node. There exist three such routes for the specified project. They are as follows:

Path I

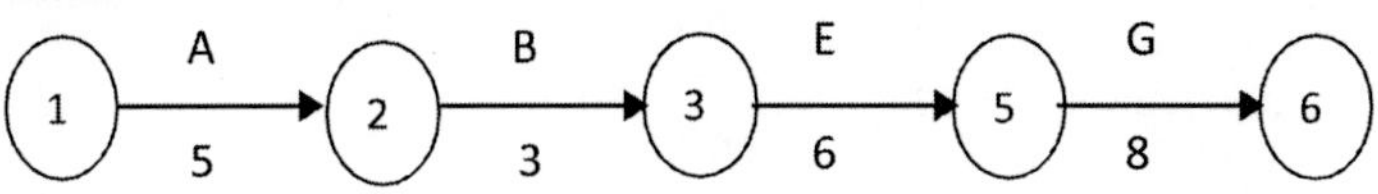

Time for the route: 5+3+6+8 = 22 weeks.

Path II

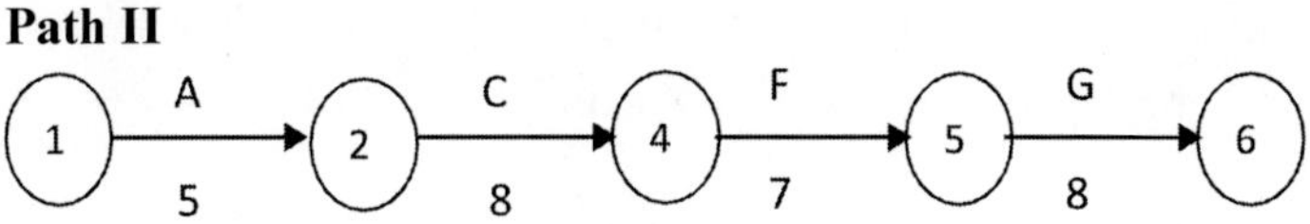

Time required for the path: 5+8+7+ 8 = 28 weeks.

Path III

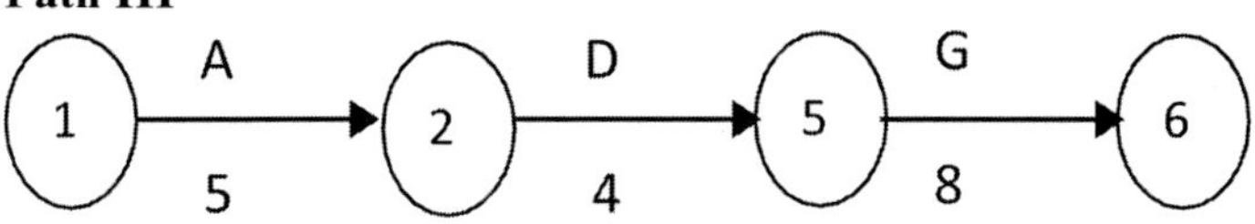

Time for the route: 5+4+8 = 17 weeks.

Compare the durations for the three routes.

Greatest of {22, 28, 17} = 28.

It is observed that Path II has the greatest time.

Thus, the critical path is Path II. i. e. , 1 2 4 5 6.

The essential activities are A, C, F, and G.

The non-essential activities are B, D, and E.

Project duration = 28 weeks.

Calculation of Standard Deviation and Variance for the Critical Activities:

Critical Activity	Optimistic time estimate (to)	Most likely time estimate (tm)	Pessimistic time estimate (tp)	Range (tp -to)	Standard deviation $\sigma = \frac{t_p - t_o}{6}$	Variance $\sigma^2 = \left(\frac{t_p - t_o}{6}\right)^2$
A: 1 2	2	5	8	6	1	1
C: 2 4	6	8	10	4	2/3	4/9
F: 4 5	6	7	8	2	1/3	1/9
G: 5 6	6	8	10	4	2/3	4/9

Standard deviation of the critical path = √2 = 1. 414

The formula for the standard normal variate is provided by

$$Z = \frac{\text{Given value of t - Anticipated Value of t in the Critical Path}}{\text{SD for the Critical Path}}$$

So we get $Z = \frac{30-28}{1.414} = 1.414$

We consult the Normal Probability Distribution Table.

For Z = 1. 414, we find the value of 0. 4207.

We calculate 0. 5 + 0. 4207 = 0. 9207.

Thus, the necessary probability is 0. 92.

That is, there is a 92% likelihood that the project will be finished before 30 weeks.

To put it differently, the probability that it will be postponed past 30 weeks is 8%.

Questions

1. Describe how the duration of an activity is calculated in PERT.
2. Describe the assessment of certainty in PERT.
3. The time estimates in weeks for the activities of a project are as follows:

Activity	Predecessor Activity	Optimistic estimate of time	Most likely estimate of time	Pessimistic estimate of time
A	-	2	4	6
B	A	8	11	20
C	A	10	15	20
D	B	12	18	24
E	C	8	13	24
F	C	4	7	16
G	D,F	14	18	28
H	E	10	12	14
I	G,H	7	10	19

Identify the essential tasks and the duration required to finish the project.

4. Create the network diagram for the subsequent project. Calculate the duration, variance, and standard deviation of the project.

Activity	Predecessor Activity	Optimistic estimate of time	Most likely estimate of time	Pessimistic estimate of time
A	-	12	14	22
B	-	16	17	24
C	A	14	15	16
D	A	13	18	23
E	B	16	18	20
F	D, E	13	14	21
G	C, F	6	8	10

5. Reflect on the subsequent project that includes the projected duration in weeks:

Activity	Predecessor Activity	Optimistic estimate of time	Most likely estimate of time	Pessimistic estimate of time
A	-	2	4	6
B	-	3	5	7
C	A	5	6	13
D	A	4	8	12
E	B,C	5	6	13
F	D,E	6	8	14

Determine the likelihood that the project will be finished in 27 weeks.

NORMAL DISTRIBUTION TABLE

Area Under Standard Normal Distribution

	0.00	0.01	0.02	0.03	0.04	0.05	0.06	0.07	0.08	0.09
0.0	0.0000	0.0040	0.0080	0.0120	0.0160	0.0199	0.0239	0.0279	0.0319	0.0359
0.1	0.0398	0.0438	0.0478	0.0517	0.0557	0.0596	0.0636	0.0675	0.0714	0.0753
0.2	0.0793	0.0832	0.0871	0.0910	0.0948	0.0987	0.1026	0.1064	0.1103	0.1141
0.3	0.1179	0.1217	0.1255	0.1293	0.1331	0.1368	0.1406	0.1443	0.1480	0.1517
0.4	0.1554	0.1591	0.1628	0.1664	0.1700	0.1736	0.1772	0.1808	0.1844	0.1879
0.5	0.1915	0.1950	0.1985	0.2019	0.2054	0.2088	0.2123	0.2157	0.2190	0.2224
0.6	0.2257	0.2291	0.2324	0.2357	0.2389	0.2422	0.2454	0.2486	0.2517	0.2549
0.7	0.2580	0.2611	0.2642	0.2673	0.2704	0.2734	0.2764	0.2794	0.2823	0.2852
0.8	0.2881	0.2910	0.2939	0.2967	0.2995	0.3023	0.3051	0.3078	0.3106	0.3133
0.9	0.3159	0.3186	0.3212	0.3238	0.3264	0.3289	0.3315	0.3340	0.3365	0.3389
1.0	0.3413	0.3438	0.3461	0.3485	0.3508	0.3531	0.3554	0.3577	0.3599	0.3621
1.1	0.3643	0.3665	0.3686	0.3708	0.3729	0.3749	0.3770	0.3790	0.3810	0.3830
1.2	0.3849	0.3869	0.3888	0.3907	0.3925	0.3944	0.3962	0.3980	0.3997	0.4015
1.3	0.4032	0.4049	0.4066	0.4082	0.4099	0.4115	0.4131	0.4147	0.4162	0.4177
1.4	0.4192	0.4207	0.4222	0.4236	0.4251	0.4265	0.4279	0.4292	0.4306	0.4319
1.5	0.4332	0.4345	0.4357	0.4370	0.4382	0.4394	0.4406	0.4418	0.4429	0.4441
1.6	0.4452	0.4463	0.4474	0.4484	0.4495	0.4505	0.4515	0.4525	0.4535	0.4545
1.7	0.4554	0.4564	0.4573	0.4582	0.4591	0.4599	0.4608	0.4616	0.4625	0.4633
1.8	0.4641	0.4649	0.4656	0.4664	0.4671	0.4678	0.4686	0.4693	0.4699	0.4706
1.9	0.4713	0.4719	0.4726	0.4732	0.4738	0.4744	0.4750	0.4756	0.4761	0.4767
2.0	0.4772	0.4778	0.4783	0.4788	0.4793	0.4798	0.4803	0.4808	0.4812	0.4817
2.1	0.4821	0.4826	0.4830	0.4834	0.4838	0.4842	0.4846	0.4850	0.4854	0.4857
2.2	0.4861	0.4864	0.4868	0.4871	0.4875	0.4878	0.4881	0.4884	0.4887	0.4890
2.3	0.4893	0.4896	0.4898	0.4901	0.4904	0.4906	0.4909	0.4911	0.4913	0.4916
2.4	0.4918	0.4920	0.4922	0.4925	0.4927	0.4929	0.4931	0.4932	0.4934	0.4936
2.5	0.4938	0.4940	0.4941	0.4943	0.4945	0.4946	0.4948	0.4949	0.4951	0.4952
2.6	0.4953	0.4955	0.4956	0.4957	0.4959	0.4960	0.4961	0.4962	0.4963	0.4964
2.7	0.4965	0.4966	0.4967	0.4968	0.4969	0.4970	0.4971	0.4972	0.4973	0.4974
2.8	0.4974	0.4975	0.4976	0.4977	0.4977	0.4978	0.4979	0.4979	0.4980	0.4981
2.9	0.4981	0.4982	0.4982	0.4983	0.4984	0.4984	0.4985	0.4985	0.4986	0.4986
3.0	0.4987	0.4987	0.4987	0.4988	0.4988	0.4989	0.4989	0.4989	0.4990	0.4990

Example 9

A project consists of seven activities. Activities P, Q, R run simultaneously.

Activity	Immediate successor
P	S
Q	T
R	U

Activity V is the final operation of the project and also serves as the direct successor to S, T, and U. Illustrate the network of the project.

Solution

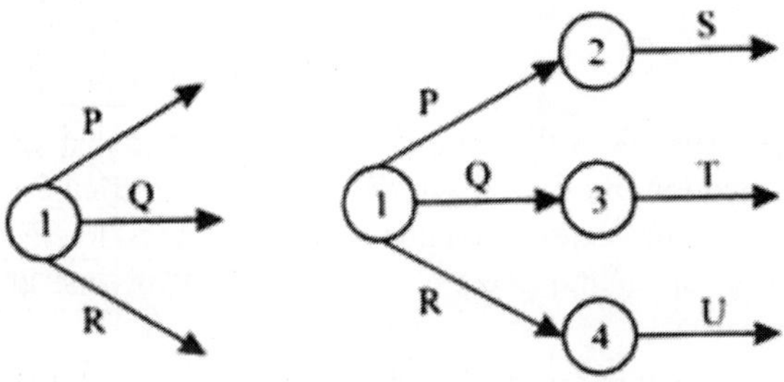

The network diagram can be developed as follows

1. Activities P, Q and R are concurrent activities initiating from node 1
2. Now since S, T and U are the immediate successors to activities P, Q and R respectively.
3. Also V is the last operation or immediate successor to S, T and U so the network becomes.

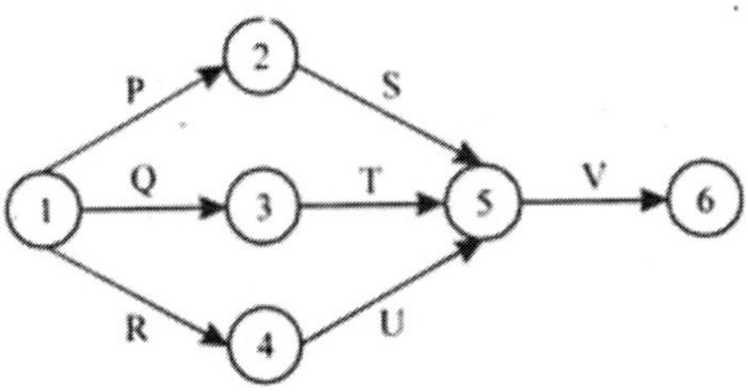

Example 10

Draw network of the Project with following situation:

i. P is prerequisite of S
ii. Q is prerequisite of S and T
iii. R is prerequisite of T
iv. S and T are prerequisites of U

Solution

These activities are shown below

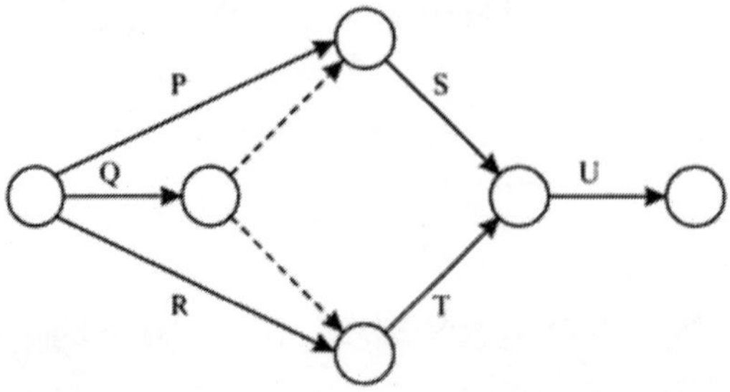

Estimates of time of activities

After presenting the actions and occurrences within the network form, after that we must establish the timing for the execution of all such activities. It is very difficult to ascertain the timing of an activity, because it depends on many factors. The following three various kinds of estimates are taken into account for the length of time of activities.

Optimistic time (to) : the least amount of time needed to finish the activity is called Optimistic time and denoted by t_o.

Pessimistic time (t_p): the maximum time needed to finish the activity is called pessimistic time and denoted by t_p.

Most likely time (tm): the duration needed to finish a task in standard circumstances is referred to as most likely time and denoted by $t_{m.}$

The anticipated duration for finishing the activity may be obtained from these three times discussed above. The anticipated duration represented by t_e obtained by the following formula

$$\text{Expected time } t_e = \frac{t_o + 4\,t_m + t_p}{6}$$

Where, te = Expected time

to = Optimistic time

tp = Pessimistic time

tm = Most likely time

Or

$$\text{Expected time } t_e \; \frac{a + 4m + b}{6}$$

Where, t_e = Expected time

a = Optimistic time assessment

m = Most probable time assessment

b = Pessimistic time assessment

variance of the project $\sigma^2 = \left(\frac{(a-b)}{6}\right)^2$ and standard deviation is $\sigma = \frac{(a-b)}{6}$

or

variance of the project $\sigma^2 = \left(\frac{t_p - t_o}{6}\right)^2$ and standard deviation is $\sigma = \frac{t_p - t_o}{6}$

Example 11

For the project's activity, the Optimistic time is 6 hours, the Pessimistic time is 10 hours, and the Most likely time is 12 hours. Determine the expected time for the activity.

Solution

$$\text{Expected time } t_e = \frac{t_o + 4\,t_m + t_p}{6}$$

Where,

$$\text{Expected time } t_e = \frac{6 + 4(12) + 10}{6}$$

$$= 10.67 \text{ hour}$$

Example 12

The estimates of times of different activities of project hours are provided below. Obtain anticipated duration of each task and draw PERT chart.

Activity	Most likely time	Optimistic time	Pessimistic time
1-2	4	3	6
2-3	8	2	10
2-4	10	4	14
3-5	8	5	13
4-5	3	2	4
4-6	7	4	8
5-7	10	6	12
6-7	3	1	5
7-8	5	3	9
7-9	3	2	4
8-10	9	7	11
9-10	7	5	13

Solution

We will derive the anticipated time using the subsequent formula

Expected time $t_e = \dfrac{t_o + 4\,t_m + t_p}{6}$

Activity	Most likely time	Optimistic time	Pessimistic time	Expected time (t_e)
1-2	4	3	6	4.17
2-3	8	2	10	7.33
2-4	10	4	14	9.67
3-5	8	5	13	8.33
4-5	3	2	4	3.00
4-6	7	4	8	6.67
5-7	10	6	12	9.67
6-7	3	1	5	3.00
7-8	5	3	9	5.33
7-9	3	2	4	3.00
8-10	9	7	11	9.00
9-10	7	5	13	7.67

From the order of the tasks of the table we shall draw the PERT diagram as follow.

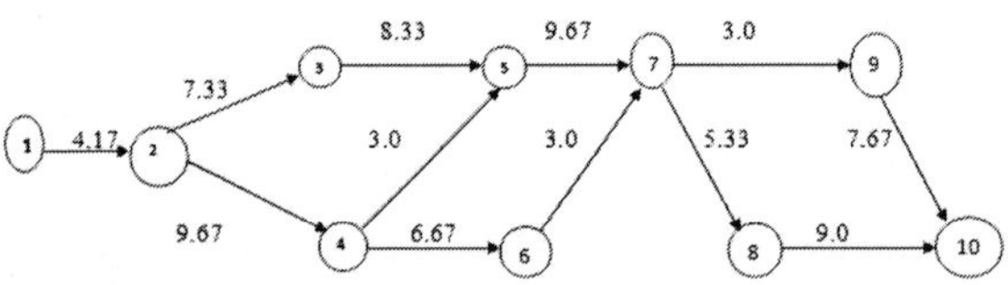

Example 13

Draw the PERT diagram as per the following table information.

Activity	1-2	1-3	2-3	2-4	3-4	4-5
Time	5	7	3	4	3	5

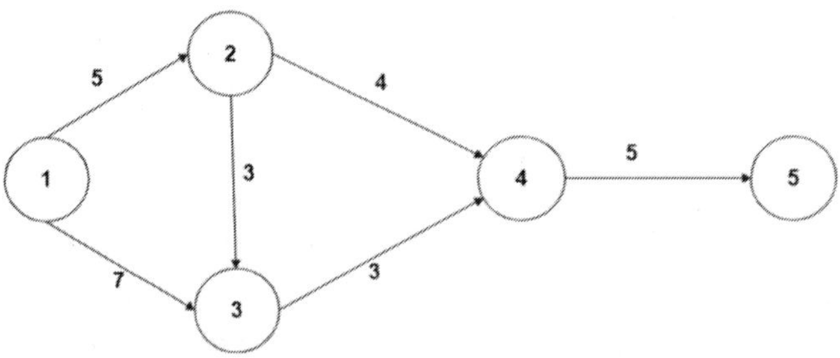

Example 14

Draw the PERT diagram as per the following table information.

Activity	1-2	1-3	2-3	2-4	3-4	4-5	3-6	5-6
Time	11	14	6	7	3	5	3	2

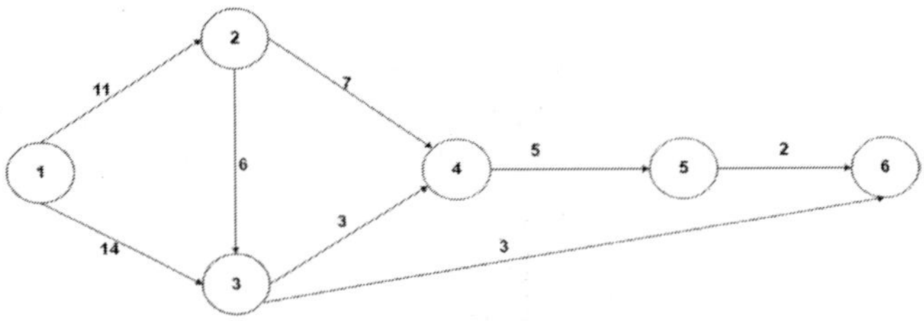

Example 15

Draw the PERT diagram as per the following table information.

Activity	1-2	1-6	2-3	2-4	3-5	4-5	6-7	5-8	7-8
Time	8	4	14	7	13	8	10	5	20

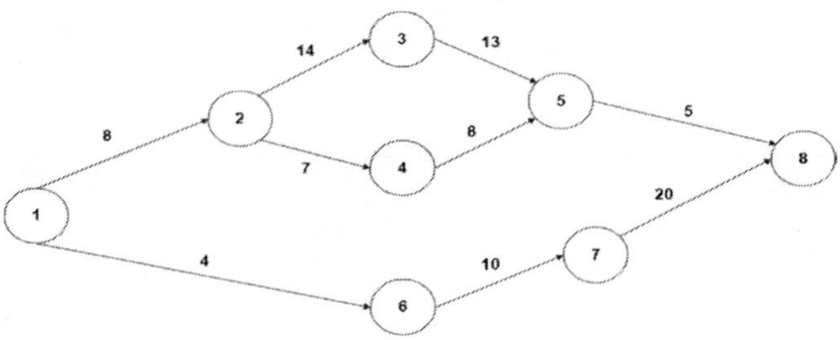

Example 16

Draw the PERT diagram as per the following table information.

Activity	Sequence	a	m	b	Float Time
a	1-2	7	12	13	11.33
b	1-3	7	10	12	9.83
c	2-5	8	13	15	12.5
d	3-5	10	12	22	13.33
c	5-6	10	14	18	14.0

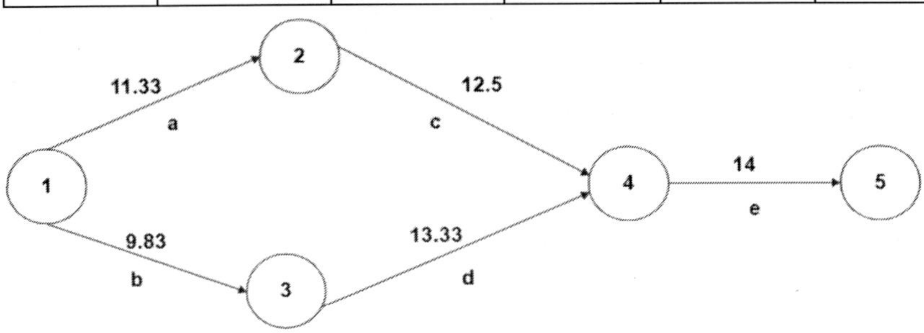

Example 17

Draw the PERT diagram as per the following table information.

Activity	Optimistic Time (Days)	Most Likely Time (Days)	Pessimistic Time (Days)	Expected Time (Days)
1-2	1	3	5	3
2-3	4	8	10	7.67
2-4	1	3	6	3.17
3-4	2	4	7	4.17
3-5	2	6	8	5.67
4-6	2	3	4	3
5-8	2	5	8	5
6-7	6	8	10	8
6-10	3	4	5	4
7-9	2	5	7	4.83
8-9	3	5	8	5.17
9-10	6	7	8	7

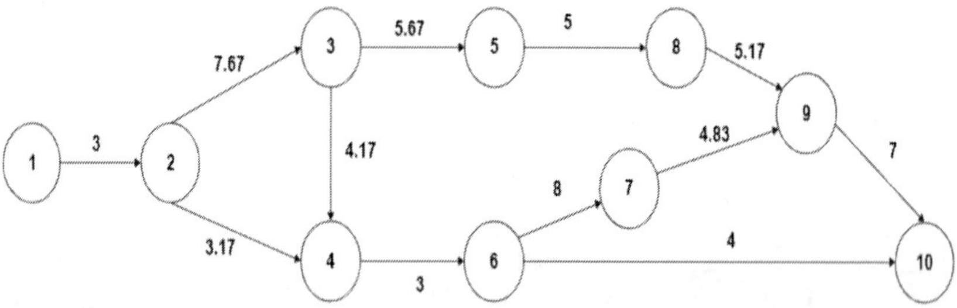

Example 18

Draw the PERT diagram as per the following table information. Find out critical path and critical time.

Activity	Preceeding Time	Time (Days)
a	-	2
b	a	3
c	b	4
d	-	5
e	d	2
f	c, e	5

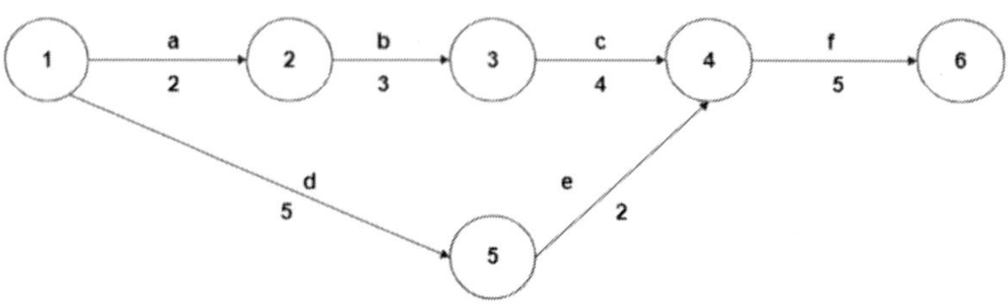

Path P1: 1-2-3-4-6 Time = 14

P2: 1-5-4-6 Time = 12

Critical path for the network **P1: 1-2-3-4-6** and total time required is 14 days

Example 19

Draw the PERT diagram as per the following table information. Find out Number of paths, critical path and critical time.

Job	a	m	b	Expected Time
1-2	1	2	3	2
1-3	1.5	2	2.5	2
1-4	2	4	9	4.5
2-5	2	3	7	3.5
3-6	4	4.5	8	5
3-7	6	8.25	9	8
4-7	3	3.5	7	4
5-8	2	2	2	2
6-8	2	4	6	4
7-9	2	5	8	5
8-9	2	3	4	3
9-10	2.5	4	5.5	4

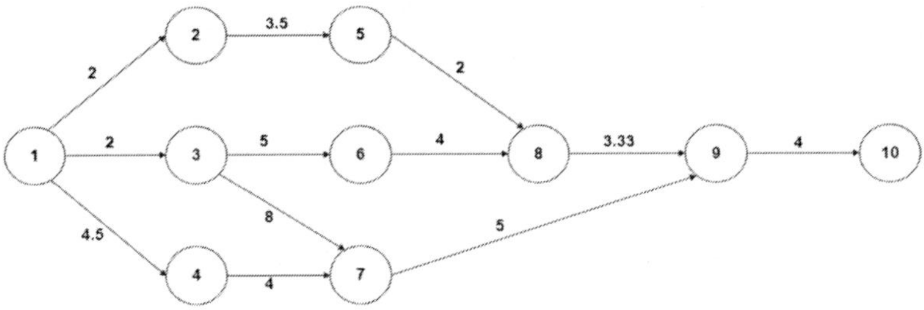

Path P1: 1-2-5-8-9-10 Time = 14.83

P2: 1-3-6-8-9-10 Time = 18.33

P3: 1-3-7-9-10 Time = 19

P4: 1-4-7-9-10 Time = 17.5

Critical path for the network **P3: 1-3-7-9-10** and total time required is **19** days

Example 20

Draw the PERT diagram as per the following table information. Find out Number of paths, critical path and critical time.

Activity	1-2	1-3	1-4	2-5	3-7	4-6	5-7	5-8	6-9	7-10	8-10	9-10
Time	12	9	10	9	16	8	10	7	6	14	15	16

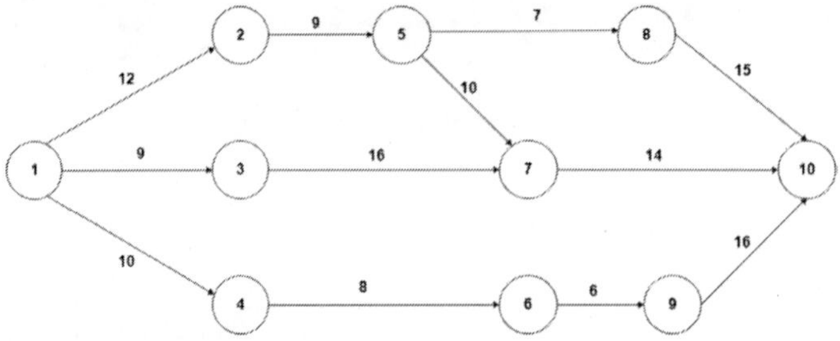

Path P1: 1-2-5-8-10 Time = 43

P2: 1-2-5-7-10 Time = 45

P3: 1-3-7-10 Time = 39

P4: 1-4-6-9-10 Time = 40

Critical path for the network **P2: 1-2-5-7-10** and total time required is **45**

Example 21

Draw the PERT diagram as per the following table information. Find out Number of paths, critical path and critical time.

Activity	1-2	1-3	1-4	2-3	2-5	3-5	4-5
Time	4	2	5	7	7	2	5

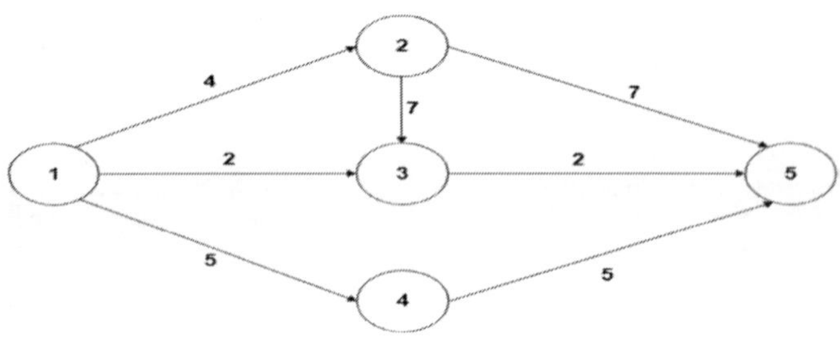

Path P1: 1-2-5 Time = 11

P2: 1-2-3-5 Time = 13

P3: 1-3-5 Time = 4

P4: 1-4-5 Time = 10

Critical path for the network P2: 1-2-3-5 and total time required is 13 hours

Example 22

Draw the PERT diagram as per the following table information. Find out Number of paths, critical path and critical time.

Activity	Time (Hours)
1-2	13
2-4	12
2-3	13
3-4	10
3-5	13
4-5	17
3-6	12
5-6	16
4-6	15

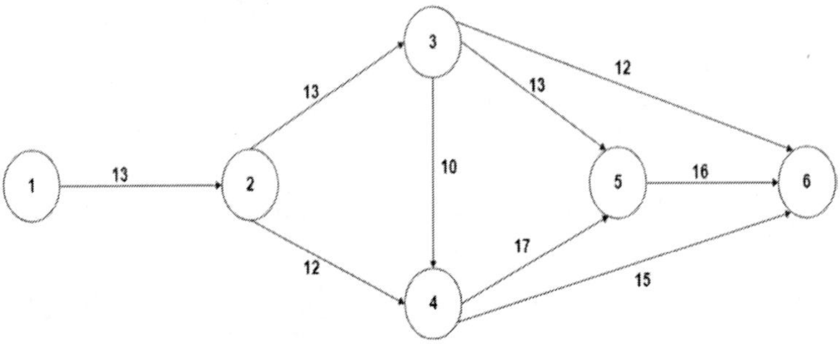

Path P1: 1-2-3-6 Time = 38

P2: 1-2-3-5-6 Time = 55

P3: 1-2-3-4-5-6 Time = 69

P4: 1-2-4-6 Time = 51

P5: 1-2-4-6 Time = 40

P6: 1-2-4-5-6 Time = 58

Critical path for the network **P3: 1-2-3-4-5-6** and total time required is **69**

Example 23

Draw the PERT diagram as per the following table information. Find out Number of paths, critical path and critical time.

Activity	Preceeding Activity	a	m	b	Expected Time
a	-	4	6	8	6
b	a	5	7	15	8
c	a	4	8	12	8
d	b	15	20	25	20
e	b	10	18	26	18
f	c	8	9	16	10

g	e	4	8	12	8
h	d,f	1	2	3	2
i	g,h	6	7	8	7

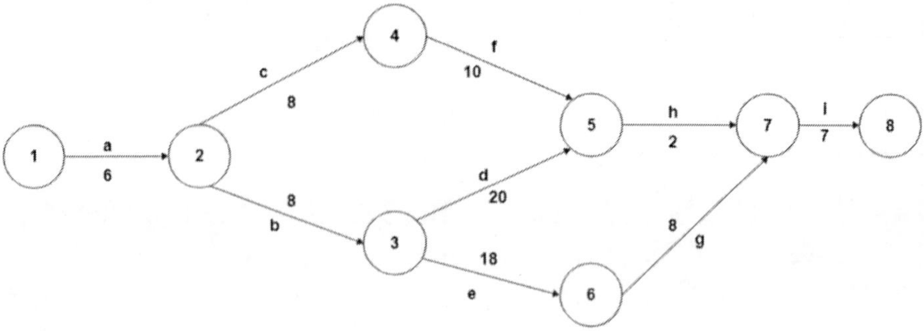

Path P1: 1-2-4-5-7-8 Time = 33

P2: 1-2-3-5-7-8 Time = 43

P3: 1-2-3-6-7-8 Time = 47

Critical path for the network **P3: 1-2-3-6-7-8** and total time required is **47** days

Example 24

Draw the PERT diagram as per the following table information. Find out critical path and critical time.

Job	Time	EST	EFT	LST	LFT	Float Time
1-2	2	0	2	2	4	2
1-3	5	0	5	0	5	0
2-4	4	2	6	4	8	2
3-4	3	5	8	5	8	0
3-5	5	5	10	11	16	6
4-6	6	8	14	8	14	0
5-7	2	10	12	16	18	6
6-7	4	14	18	14	18	0

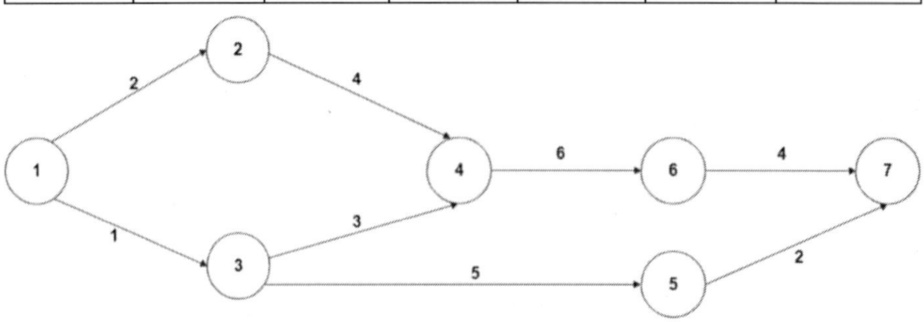

Critical path for the network **1-3-4-6-7** and total time required is **18**

Example 25

Draw the PERT diagram as per the following table information. Find out critical path and critical time.

Activity	Days	EST	EFT	LST	LFT	Float Time
1-2	20	0	20	0	20	0
1-3	28	0	25	5	30	5
2-3	10	20	30	20	30	0
2-4	12	20	32	23	35	3
3-4	5	30	35	30	35	0
4-5	10	35	45	30	45	0

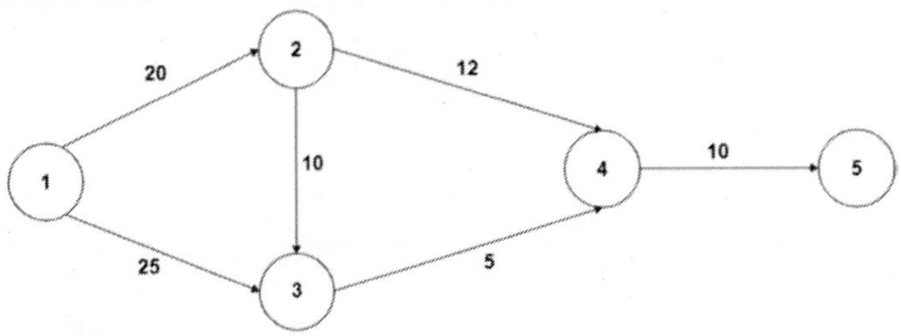

Critical path for the network **1-2-3-4-5** and total time required is **45** days

Example 26

Draw the PERT diagram as per the following table information. Find out Number of paths, critical path and critical time

Activity	Time
1-2	13
2-3	13
2-4	12
3-4	10
3-5	13
4-5	17
4-6	15
5-6	16

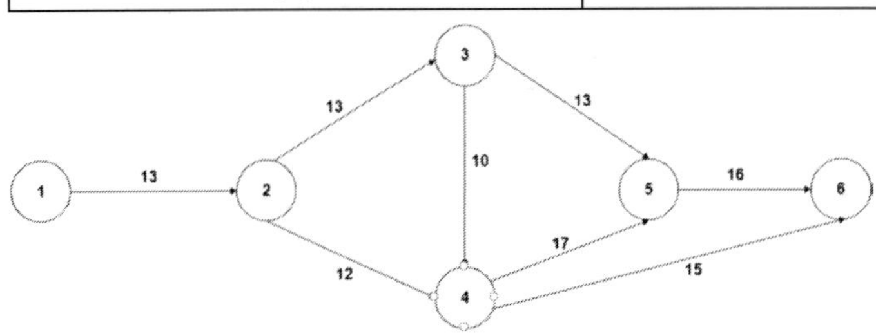

Path P1: 1-2-3-5-6 Time = 55

P2: 1-2-3-4-5-6 Time = 69

P3: 1-2-4-5-6 Time = 58

P4: 1-2-4-6 Time = 40

Critical path for the network **P2 1-2-3-4-5-6** and total time required is **69** hours

Example 27

Draw the PERT diagram as per the following information according to figure no 3.4

Activity	Time (Hours)	EST	EFT	LST	LFT	Float Time
1-2	2	0	2	2	4	2
1-3	5	0	5	5	5	0
2-4	4	2	6	4	8	2
3-4	3	5	8	5	8	0
3-5	5	5	10	9	14	4
4-5	6	8	14	8	14	0
5-6	2	14	16	14	16	0

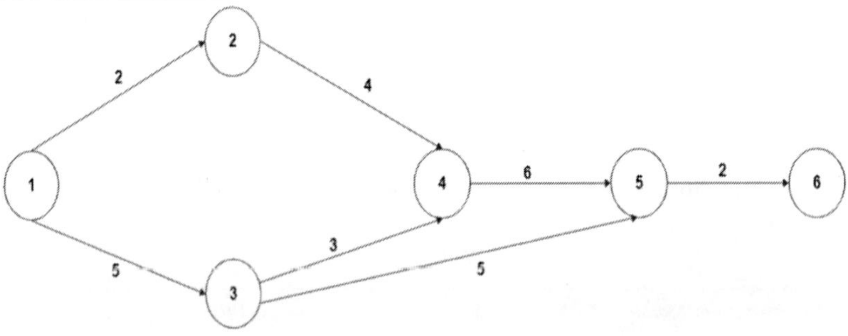

Example 28

Draw the PERT diagram as per the following table information. Find out critical path, critical time, variance of the project, and Z value.

Activity	Time	EST	EFT	LST	LFT	Slack time	Variance
1-2	15	0	15	0	15	0	2.78
1-6	8.83	0	8.83	64.51	73.34	64.51	1.36
2-3	24.17	15	39.17	15	39.17	0	6.25
3-4	19.5	39.17	58.67	39.17	58.67	0	4.70
3-5	8.17	39.17	47.39	51.5	59.67	12.33	1.36
4-7	25	58.67	83.67	58.67	83.67	0	11.10
4-8	14.67	58.67	73.34	78.17	92.84	19.5	4
4-9	23	58.67	81.67	61.14	84.84	3.17	5.44
5-6	13.67	47.34	61.02	59.67	73.34	12.32	5.44
6-10	20.17	61.01	81.18	73.34	93.51	12.33	2.25

7-11	7.17	83.67	90.84	88.67	90.84	0	2.25
8-10	0	73.34	73.34	93.51	93.51	20.17	0
8-12	5	73.34	78.34	92.84	97.84	19.5	0.45
9-10	8.67	81.67	90.84	84.84	93.51	3.17	1.77
10-12	4.33	90.34	94.67	95.51	97.84	3.17	1
10-13	5.55	90.34	95.84	113.17	118.64	22.83	0.69
11-12	7	90.34	97.84	90.87	97.84	0	1
12-13	20.83	97.84	118.57	97.84	118.67	0	6.25

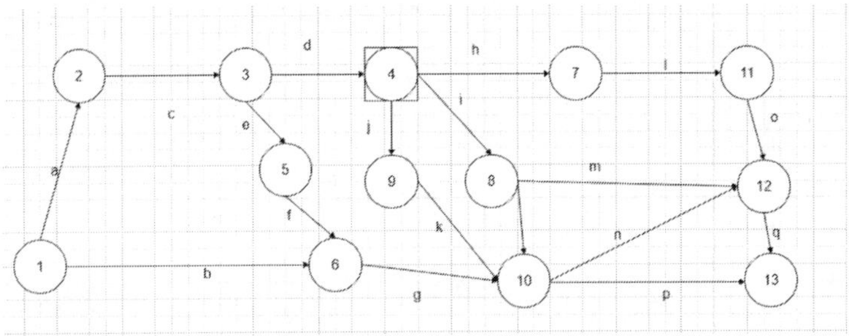

Critical path 1-2-3-4-7-11-12-13 = 118.67

Variance of project = 5.86

Z = 120-118.67/5.86 = 0.23

Example 29

Draw the PERT diagram as per the following table information. Find out Number of paths, and time.

Activity	Activity time		Activity cost						
	Normal	Crash	Normal	Crash	Total allowable crash time	Per week crash course	Cost diffe-rence	Time red-uced	cost
1-2	20	8	1000	1480	12	40	480	10	400
1-4	24	20	1200	1400	4	50	200	4	200
1-3	14	7	700	1190	7	70	490	-	-
2-4	10	6	500	820	4	80	320	-	-
3-4	11	5	550	730	6	30	180	5	150
			3950	5620					750

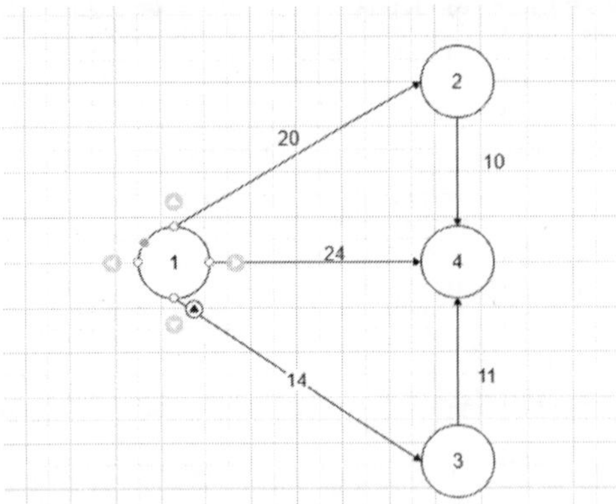

Path 1-2-4 = 30-10 = 20

1-4 = 24 – 4 = 20

1 -3-4 = 25-5 = 20

Example 30

Draw the PERT diagram as per the following table information. Find out Number of paths, and time. Also crash the network maximum amount possible.

Activity	Normal time (week)	Crash Time (week)	Normal Cost (Rs.)	Crash Cost (Rs.)
1-2	16	8	2000	4400
1-3	14	9	1000	1800
2-4	8	6	500	700
2-5	5	4	600	1300
3-5	4	2	1500	3000
3-6	6	4	800	1600
4-6	10	7	3000	4500
5-6	15	10	5000	8000

Solution

Activity	Normal	Crash	Available time	Cost difference	Cost per week	Time reduced	cost
1-2	16	8	8	2400	300	8	2400
1-3	14	9	5	800	160	5	800
2-4	8	6	2	200	100	2	200
2-5	5	4	1	700	700	1	700
3-5	4	2	2	1500	750	1	750
3-6	6	4	2	800	400	0	-
4-6	10	7	3	1500	500	2	1000
5-6	15	10	5	3000	600	5	3000
							8850

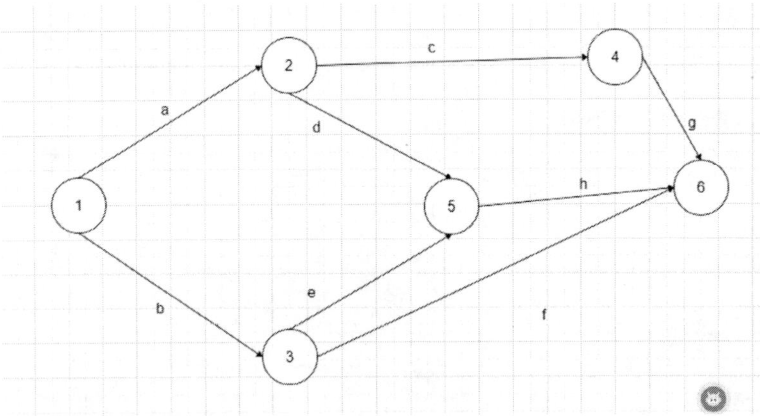

P1 1-2-4-6 = 34-12 = 22

Path P2 is Normal critical path 1-2-5-6 and normal critical path time is 36

P2 1-2-5-6 = 36-14 = 22

P3 1-3-5-6 =33-11 = 21

P4 1-3-6 =20-5= 15

Minimum project completion time is 22 and total cost is 8850.

Example 31

Draw the PERT diagram as per the following table information. Find out Number of paths, and time. Also crash the network maximum amount possible and indicate the total crash cost.

Activity	Normal Time (Weeks)	Crash Time (Weeks)	Normal Cost (Rs.)	Crash Cost (Rs.)
1-2	8	5	100	400
1-3	10	7	250	400
2-3	5	3	400	800
2-4	3	1	200	400
3-4	6	4	150	300
3-5	3	3	100	100
4-5	4	3	300	500

Solution

Activity	Normal Time (Weeks)	Crash Time (Weeks)	Available time	Cost difference	Cost per week	Time reduced	Cost
1-2	8	5	3	300	100	3	300
1-3	10	7	3	150	50	2	100
2-3	5	3	2	400	200	2	400
2-4	3	1	2	200	100	0	0
3-4	6	4	2	150	75	2	150
3-5	3	3	0	0	0	0	0
4-5	4	3	1	200	200	1	200
							1150

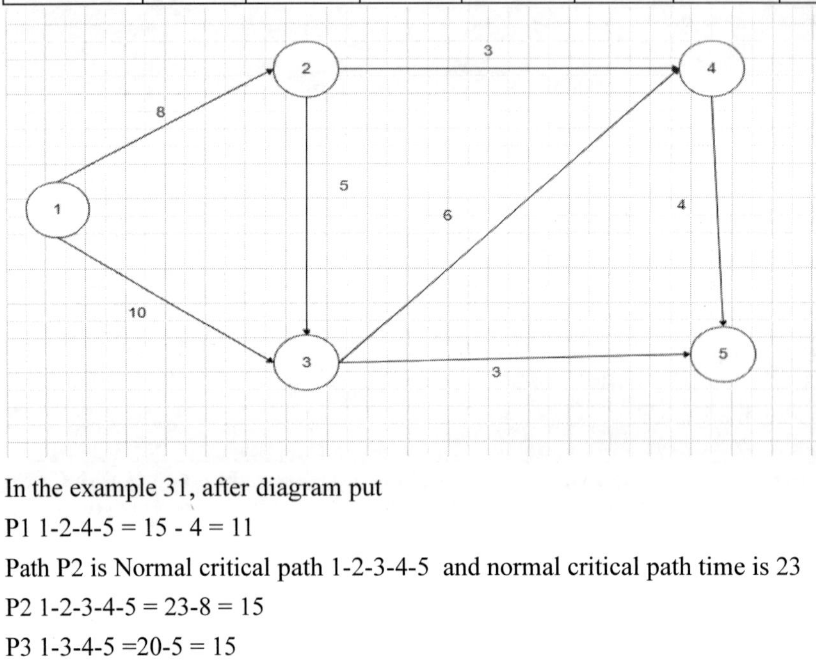

In the example 31, after diagram put

P1 1-2-4-5 = 15 - 4 = 11

Path P2 is Normal critical path 1-2-3-4-5 and normal critical path time is 23

P2 1-2-3-4-5 = 23-8 = 15

P3 1-3-4-5 =20-5 = 15

P4 1-3-5 =13-2= 11

Minimum project completion time is 15, and total cost is 1150.

Example 32

Draw the PERT diagram as per the following table information. Find out EST, EFT, LST, LFT, float time, and variance.

Activity	Time	EST	EFT	LST	LFT	Slack time	Variance
1-2	7.17	0	7.17	0	7.17	0	0.69
1-3	18	0	18	19.17	37.17	19.17	7.11
1-4	21.17	0	21.17	15.47	36.64	15.47	12.25

1-6	8	0	8	29.17	37.17	29.17	1.36
2-5	30	7.17	37.17	7.17	37.17	0	0
3-6	0	18	18	37.17	37.17	19.17	0
3-9	21.33	18	39.33	37.01	58.34	19.01	7.11
3-10	20.17	18	38.17	33.67	53.84	15.67	3.36
4-6	0	21.17	21.17	37.67	37.67	16.5	0
5-6	0	37.17	37.17	37.67	37.67	0	0
5-12	10.17	37.17	47.34	56.5	66.67	3.33	1.36
5-7	8.67	37.17	45.84	37.17	45.84	0	1.78
6-7	7	45.84	52.84	45.84	52.84	0	1
7-8	7.33	45.84	53.17	58.84	66.17	13	1.78
7-11	27.83	45.84	73.67	65	92.83	19.16	6.25
7-16	13.33	52.84	66.17	52.84	66.17	0	4
8-11	7.83	39.23	47.06	58.84	66.67	19.61	0.7
9-11	12.33	38.17	50.5	53.84	66.17	15.67	1.77
10-11	18.33	66.17	84.5	66.17	84.5	0	2.78
11-13	0	66.17	66.17	84.16	84.16	17.99	0
11-14	7	66.17	73.17	85.33	92.33	19.16	1
12-15	17.5	47.34	64.84	66.66	84.16	19.32	4.7
13-16	8.33	84.5	92.83	84.5	92.83	0	0.7
14-15	8.17	66.17	74.34	84.16	92.33	17.99	1.36
15-16	8.17	64.87	73.04	84.16	92.33	19.29	1.36

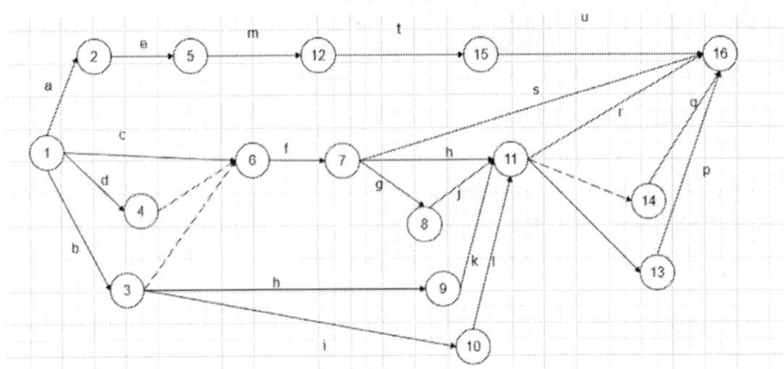

Critical path 1-2-5-6-7-8-11-13-16 = 72-88

Variance of project = 3.31

Z = 1.864

Earliest and Latest Times

Lesson Outline

- The Ideas Of Earliest And Latest Moments
- The Notion Of Slack
- Mathematical Problems

Learning Objectives

- Upon completing this lesson, you ought to be capable of
- Grasping the ideas of earliest and latest times
- Comprehending the notion of slack
- Determining the earliest and latest times
- Discovering the slacks
- Recognizing the critical activities
- Solving numerical problems

Introduction

A project manager is tasked with ensuring that a project is finished by the designated deadline, without any setbacks. This aspect will be emphasized in the subsequent discussion.

Key concepts

Certain important concepts are presented below.

Earliest Times of An Activity

We can think about (i) the Earliest Start Time of an activity and (ii) the Earliest Finish Time of an activity.

Earliest Start Time of an activity refers to the soonest time that activity can commence, provided that all preceding activities were started at their earliest possible times.

Earliest Finish Time of an activity refers to the earliest possible moment for finishing that specific activity. It is calculated using the formula.

The Earliest Finish Time of an activity = The Earliest Start Time of the activity + The projected duration for completing that activity.

Latest Times of An Activity

We may examine (i) the Latest Finish Time of a task and (ii) the Latest Start Time of a task.

Latest Finish Time of an activity is the final possible time for completing that activity provided that all subsequent activities are executed according to the management's plan and without causing any delay to the project beyond the specified time.

Latest Start Time of an activity is the most recent potential time to commence that activity. It is determined by the formula

Latest Start Time of an activity = The Latest Finish Time of the activity - The projected duration needed to complete that activity.

Total Float of an Activity

Float aims to evaluate the level of delay that is permissible. It establishes a control limit for the delay.

The overall float of a task is the duration by which that task can be postponed without causing a delay to the entire project. It is expressed using the formula

Total Float of an Activity = Latest Finish Time of the activity - Earliest Finish Time of that activity.

It is similarly provided by the formula

Total Float of an Activity = Latest Start Time of the activity – Earliest

Start Time of that activity.

Since any postponement in a crucial task will result in a hold-up of the entire project, the overall float of a critical task must be zero.

Expected Times of An Event

An event takes place at a specific moment. We can evaluate (i) the Earliest Expected Time for the Occurrence of an event and (ii) the Latest Allowable Time for the Occurrence of an event.

The earliest anticipated time for the occurrence of an event is the soonest time one can expect that event to take place, provided that all prior activities have been finished.

The Latest Allowable Time of Occurrence of an event refers to the most recent time at which one can anticipate the event occurring without causing the project to extend beyond the designated timeframe.

Procedure To Determine the Earliest Anticipated Time of An Event

Step 1: Consider the Earliest Expected Time of Occurrence of the Start Event as zero.

Step 2: For an event that is not the Start Event, identify all routes in the network that link the Start node to the node that signifies the event being evaluated.

Step 3: In the "Forward Pass" (i.e., progression through the network from left to right), determine the total of the time durations for the activities in each path recognized in Step 2.

Step 4: The route that takes the most time in Step 3 provides the Earliest Expected Time of Occurrence for the event.

Working Rule for determining the earliest anticipated time of an event:

For an event being examined, find all the preceding events and determine their earliest anticipated times. Using the earliest anticipated time of each event, include the time duration of the activity linking that event to the event being examined. The highest value among all these figures provides the Earliest Expected Time of Occurrence for the event.

Procedure to Determine the Most Recent Permissible Time of an Event

We regard the "Backward Pass" (i. e. , traversing the network from right to left).

The most recent permissible time for the End Node to occur must coincide with the project's completion time. As a result, it should be equal to the critical path time of the project.

Step 1: Determine the most recent permissible time for the End Node to occur.

Step 2: For an event that is not the End Event, identify all routes in the network that link the End node with the node that signifies the event being evaluated.

Step 3: In the "Backward Pass" (i.e., progressing through the network from right to left), deduct the time lengths of the tasks along each of these paths.

Step 4: The Latest Allowable Time of Occurrence of the event is identified by the path that has the greatest duration in Step 3. In simpler terms, the minimum time value acquired in Step 3 provides the Latest Allowable Time of Occurrence of the event.

Working Guideline for determining the most recent permissible time of an event:

For an event being evaluated, find all the succeeding events and determine their latest permissible times. From the latest permissible time of each succeeding event, deduct the time duration of the activity that starts with the event being evaluated. The smallest value among all these results provides the Latest Permissible Time of Occurrence of the event.

Slack of an Event

The permissible duration between the happening of an event is referred to as the slack of that event. It is represented by the equation

Slack of an event = Latest Permissible Time of Occurrence of the event - Earliest Anticipated Time of Occurrence of that event.

Slack of an Activity

The slack of an activity refers to the float of the activity.

Example 33

The subsequent information is accessible concerning a project:

Activity	Predecessor Activity	Duration (Weeks)
A	-	12
B	A	7
C	A	11
D	A	8
E	A	6
F	B	10
G	C	9
H	D, F	14
I	E, G	13
J	H, I	16

Determine the earliest and latest moments, the overall float for every activity, the essential activities, and the time required for project completion.

Solution

With the supplied information, we create the subsequent network diagram for the project.

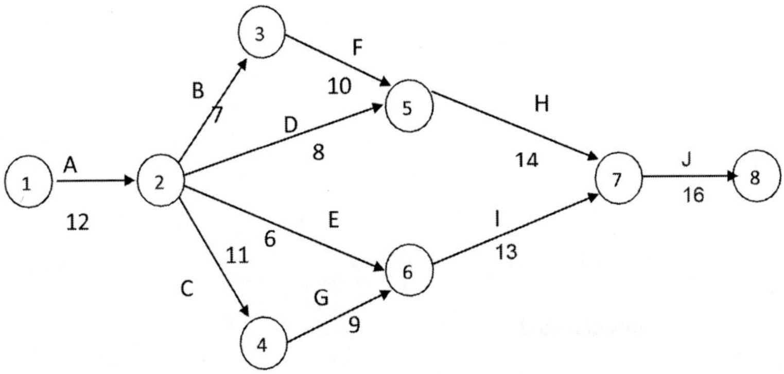

Think about the routes, initiating from the starting node and concluding with the end node. There are four of these routes for the specified project. They are as follows:

Path I

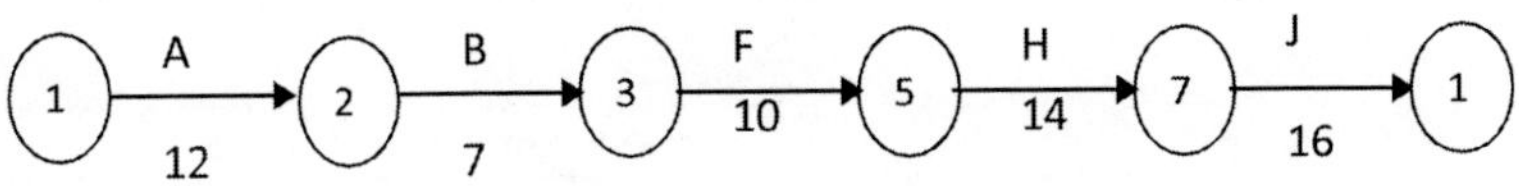

Time of the route = 12 + 7 + 10 + 14 + 16 = 59 weeks.

Path II

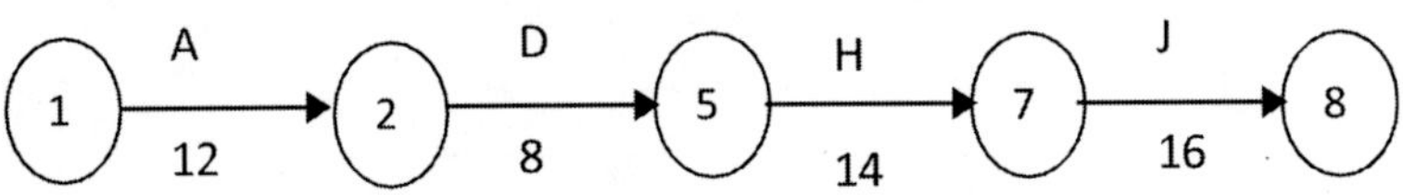

Duration of the route = 12 + 8 + 14 + 16 = 50 weeks.

Path III

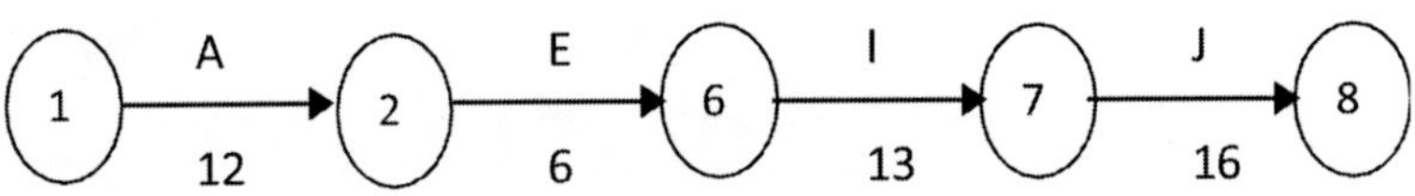

Time of the route = 12 + 6 + 13 + 16 = 47 weeks.

Path IV

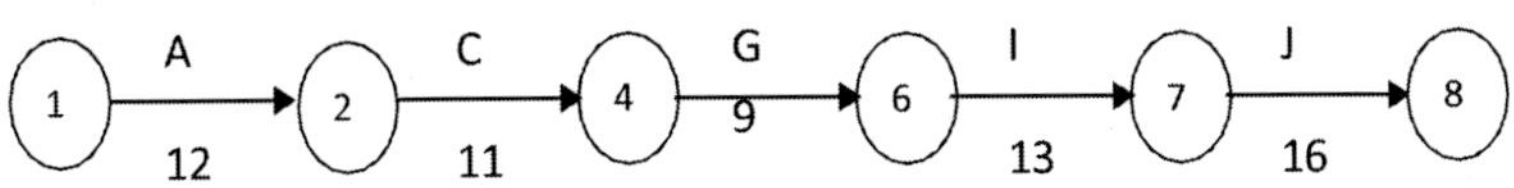

Time of the route = 12 + 11 + 9 + 13 + 16 = 61 weeks.

Compare the durations for the four routes. The highest of {51, 50, 47, 61} = 61. We observe that the longest time of a route is 61 weeks.

Forward pass

Calculation of Earliest Anticipated Time of Occurrence of Events

Node	Earliest Time of Occurrence of Node
1	0
2	Time for Node 1 + Time for Activity A = 0 + 12 = 12
3	Time for Node 2 + Time for Activity B = 12 + 7 = 19
4	Time for Node 2 + Time for Activity C = 12 + 11 = 23
5	Max {Time for Node 2 + Time for Activity D, Time for Node 3+Time for Activity F} = Max {12 + 8, 19 + 10} = Max {20, 29} = 29
6	Max {Time for Node 2 + Time for Activity E, Time for Node 4+Time for Activity G} = Max {12 + 6, 23 + 9} = Max {18, 32} = 32
7	Max {Time for Node 5 + Time for Activity H, Time for Node 6+Time for Activity I} = Max {29 + 14, 32 + 13} = Max {43, 45} = 45
8	Time for Node 7 + Time for Activity J = 45 + 16 = 61

By applying the aforementioned values, we derive the Earliest Start Times for the activities as shown below:

Activity	Earliest Start Time (Weeks)
A	0
B	12
C	12
D	12
E	12
F	19
G	23
H	29
I	32
J	45

Backward pass

Calculation of Most Recent Permissible Time of Event Occurrence

Node	Latest Allowable Time of Occurrence of Node
8	Maximum time of a path in the network = 61
7	Time for Node 8 - Time for Activity J = 61 -16 = 45
6	Time for Node 7 - Time for Activity I = 45 -13 = 32
5	Time for Node 7 - Time for Activity H = 45 -14 = 31
4	Time for Node 6 - Time for Activity G = 32 - 9 = 23
3	Time for Node 5 - Time for Activity F = 31- 10 = 21
2	Min {Time for Node 3 - Time for Activity B, Time for Node 4 - Time for Activity C, Time for Node 5 - Time for Activity D, Time for Node 6 - Time for Activity E} = Min {21 - 7, 23 - 11, 31 - 8, 32 - 6} = Min {14, 12, 23, 26} = 12
1	Time for Node 2 - Time for Activity A = 12- 12 = 0

Using the values mentioned above, we derive the Latest Finish Times for the activities as detailed below:

Activity	Latest Finish Time (Weeks)
J	61
I	45
H	45
G	32
F	31
E	32
D	31
C	23
B	21
A	12

Calculation of Overall Float for every task:

Activity	Duration (Weeks)	Earliest Start Time	Earliest Finish Time	Latest Start Time	Latest Finish Time	Total Float = Latest Finish Time - Earliest Finish Time
A	12	0	12	0	12	0
B	7	12	19	14	21	2
C	11	12	23	12	23	0
D	8	12	20	23	31	11
E	6	12	18	26	32	14
F	10	19	29	21	31	2
G	9	23	32	23	32	0
H	14	29	43	31	45	2
I	13	32	45	32	45	0
J	16	45	61	45	61	0

The activities that have a total float of 0 are A, C, G, I, and J. These activities are critical.

Project finish duration = 61 weeks.

Example 34

The subsequent information pertains to the activities within a project:

Activity	Predecessor Activity	Duration (Weeks)
A	-	15
B	A	17
C	A	21
D	B	19
E	B	22
F	C, D	18
G	E, F	15

Calculate the earliest and latest times, the total float for each task and the project's completion time.

Solution

The subsequent network diagram is acquired for the specified project.

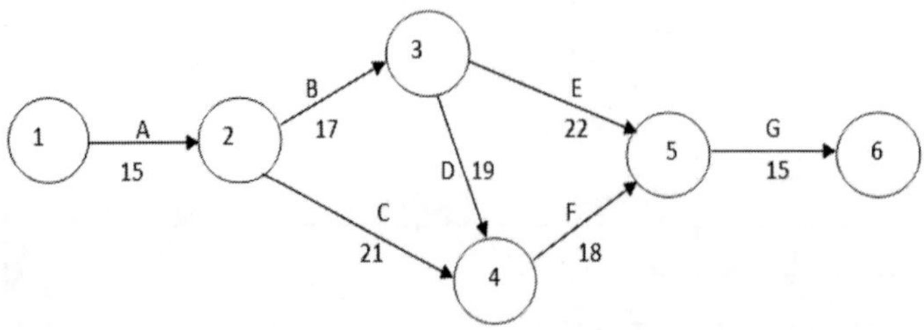

Consider the routes, starting with the initial node and concluding with the final node. There are three such routes for the specified project. They are as follows:

Path I

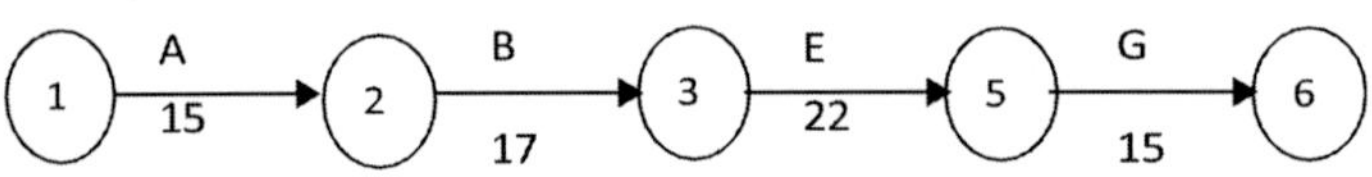

Time of the route = 15 + 17 + 22 + 15 = 69 weeks.

Path II

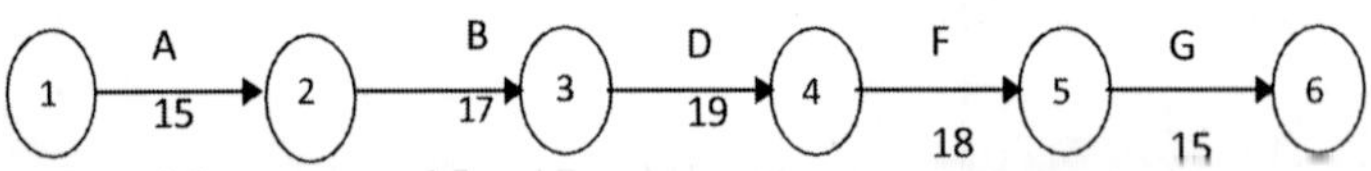

Time of the route = 15 + 17 + 19 + 18 + 15 = 84 weeks.

Path III

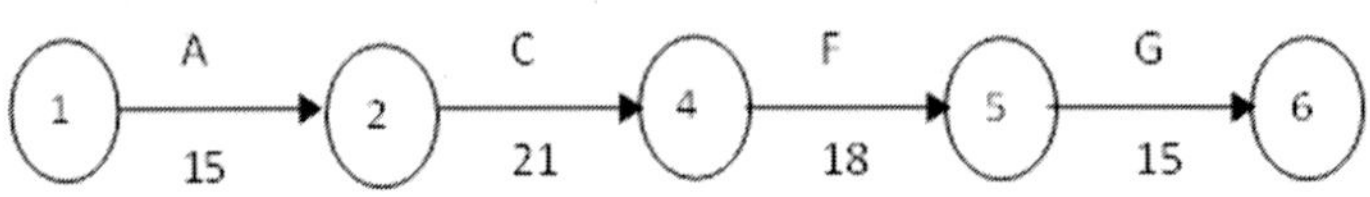

Duration of the pathway = 15 + 21 + 18 + 15 = 69 weeks.

Compare the durations for the three routes. The highest of {69, 84, 69} = 84. We observe that the longest duration of a route is 84 weeks.

Forward pass

Calculation of Earliest Time of Occurrence of Events

Node	Earliest Time of Occurrence of Node
1	0
2	Time for Node 1 + Time for Activity A = 0 + 15 = 15
3	Time for Node 2 + Time for Activity B = 15 + 17 = 32

4	Max {Time for Node 2 + Time for Activity C, Time for Node 3+Time for Activity D} = Max {15 + 21, 32 + 19} = Max {36, 51} = 51
5	Max {Time for Node 3 + Time for Activity E, Time for Node 4+Time for Activity F} = Max {32 + 22, 51 + 18} = Max {54, 69} = 69
6	Time for Node 5 + Time for Activity G = 69 + 15 = 84

Calculation of Earliest Time for Activities

Activity	Earliest Start Time (Weeks)
A	0
B	15
C	15
D	32
E	32
F	51
G	69

Backward pass

Calculation of the Most Recent Permissible Timing of Event Occurrences

Node	Latest Allowable Time of Occurrence of Node
6	Maximum time of a path in the network = 84
5	Time for Node 6 - Time for Activity G = 84 -15 = 69
4	Time for Node 5 - Time for Activity F = 69 - 18 = 51
3	Min {Time for Node 4 - Time for Activity D, Time for Node 5 - Time for Activity E} = Min {51 - 19, 69 - 22} = Min {32, 47} = 32
2	Min {Time for Node 3 - Time for Activity B, Time for Node 4 - Time for Activity C} = Min {32 - 17, 51 - 21} = Min {15, 30} = 15
1	Time for Node 2 - Time for Activity A = 15 - 15 = 0

Calculation of the Most Recent Completion Times of the activities

Activity	Latest Finish Time (Weeks)
G	84
F	69
E	69
D	51
C	51
B	32
A	15

Calculation of Total Float for every activity:

Activity	Duration (Weeks)	Earliest Start Time	Earliest Finish Time	Latest Start Time	Latest Finish Time	Total Float = Latest Finish Time - Earliest Finish Time
A	15	0	15	0	15	0
B	17	15	32	15	32	0
C	21	15	36	30	51	15
D	19	32	51	32	51	0
E	22	32	54	47	69	15
F	18	51	69	51	69	0
G	15	69	84	69	84	0

The tasks that have a total float of 0 are A, B, D, F, and G. These tasks are classified as critical activities.

The duration for project completion is 84 weeks.

Example 35

Consider a project with the subsequent details:

Name of Activity	Predecessor Activity	Duration (Weeks)
A	-	8
B	A	13
C	A	9
D	A	12
E	B	14
F	B	8
G	D	7
H	C, F, G	12
I	C, F, G	9
J	E, H	10
K	I, J	7

Establish the earliest and latest times, the total float for every activity, the critical activities, the slacks for the events, and the overall project completion time.

Solution

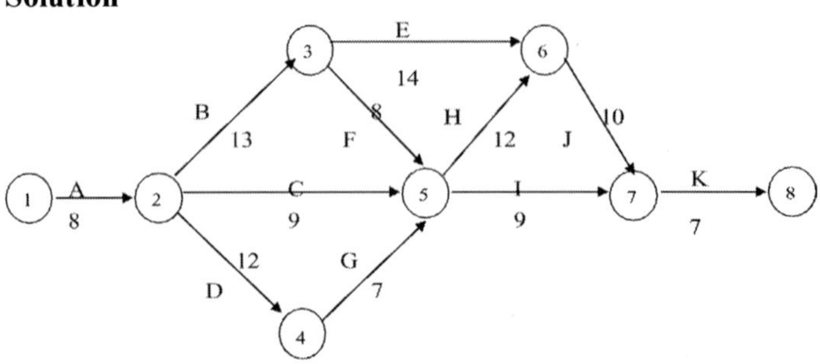

The network diagram below is obtained for the specified project:

Path I

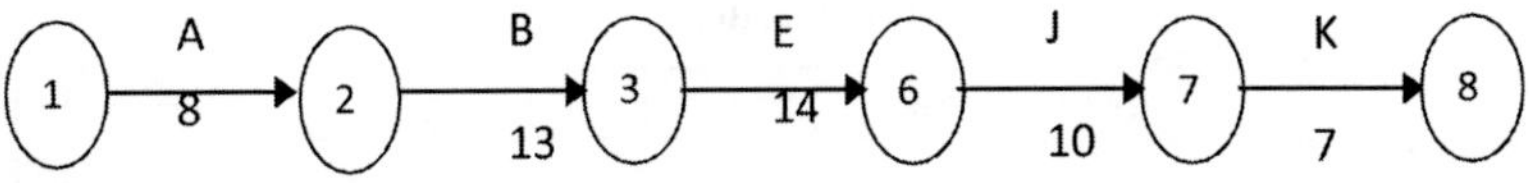

Time of the route = 8 + 13 + 14 + 10 + 7 = 52 weeks.

Path II

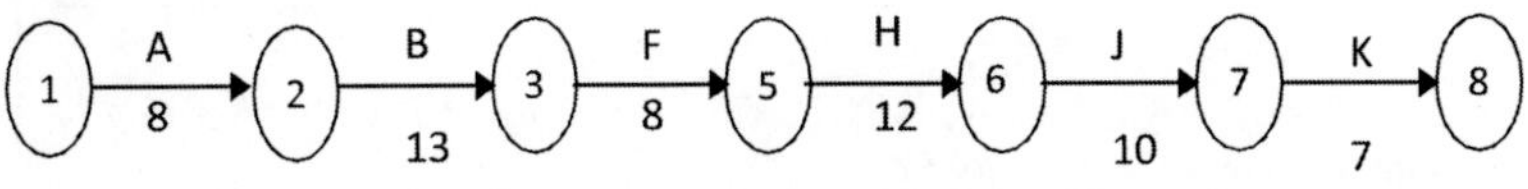

Duration of the path = 8 + 13 + 8 + 12 + 10 + 7 = 58 weeks.

Path III

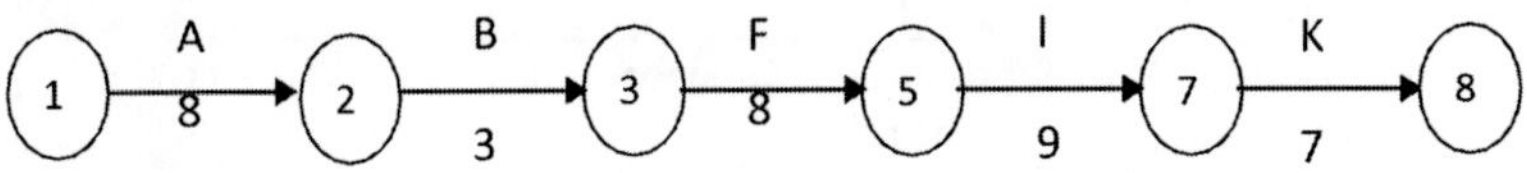

Time of the route = 8 + 13 + 8 + 9 + 7 = 45 weeks.

Path IV

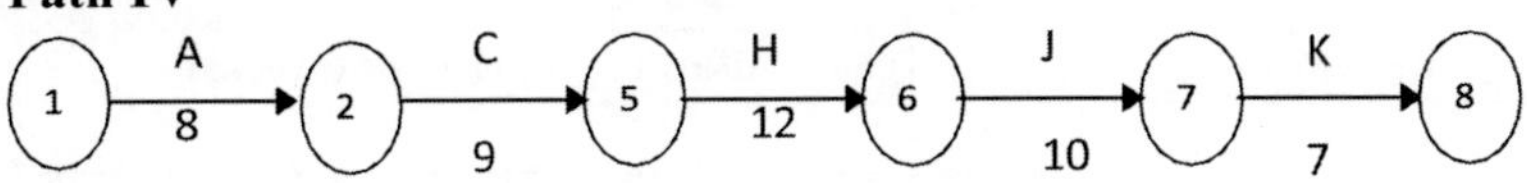

Time of the route = 8 + 9 + 12 + 10 + 7 = 46 weeks.

Path V

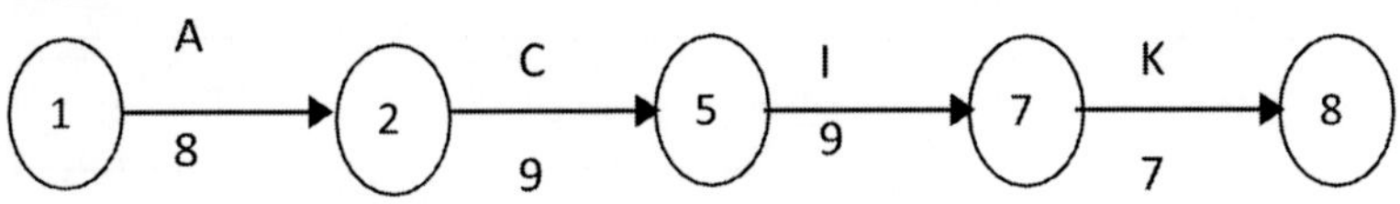

Duration of the route = 8 + 9 + 9 + 7 = 33 weeks.

Path VI

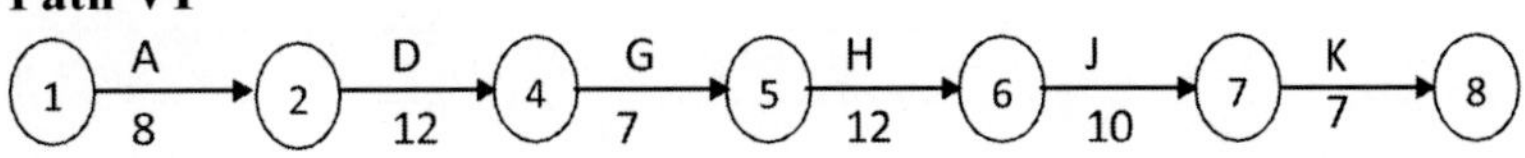

Duration of the route = 8 + 12 + 7 + 12 + 10 + 7 = 56 weeks.

Path VII

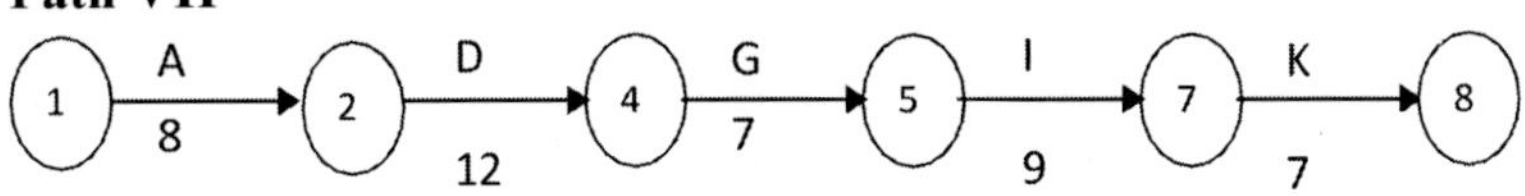

Time of the route = 8 + 12 + 7 + 9 + 7 = 43 weeks.

Compare the durations for the three routes. The highest of {52, 58, 45, 46, 33, 56, 43} = 58. We observe that the greatest time for a route is 58 weeks.

Forward pass

Calculation of Earliest Time of Occurrence of Events

Node	Earliest Time of Occurrence of Node
1	0
2	Time for Node 1 + Time for Activity A = 0 + 8 = 8
3	Time for Node 2 + Time for Activity B = 8 + 13 = 21
4	Time for Node 2 + Time for Activity D = 8+ 12 = 20
5	Max {Time for Node 2 + Time for Activity C, Time for Node 3+Time for Activity F, Time for Node 4 + Time for Activity G} = Max { 8 + 9, 21 + 8 , 20 + 7 } = Max {17, 29, 27} = 29
6	Max {Time for Node 3 + Time for Activity E, Time for Node 5+Time for Activity H} = Max {21 + 14 , 29 + 12} = Max {35, 41} = 41
7	Max {Time for Node 5 + Time for Activity I, Time for Node 6 + Time for Activity J} = Max {29 + 9, 41 + 10} = Max {38, 51} = 51
8	Time for Node 7 + Time for Activity J = 51+ 7 = 58

Earliest Start Times of the activities

Activity	Earliest Start Time (Weeks)
A	0
B	8
C	8
D	8
E	21
F	21
G	20
H	29
I	29
J	41
K	51

Backward pass

Calculation of the Most Recent Permissible Timing of Event Occurrences

Node	Latest Allowable Time of Occurrence of Node
8	Maximum time of a path in the network = 58
7	Time for Node 8 - Time for Activity K = 58 -7 = 51
6	Time for Node 7 - Time for Activity J = 51 -10 = 41
5	Min {Time for Node 6 - Time for Activity H, Time for Node 7 - Time for Activity I} = Min {41 - 12, 51 - 9} = Min {29, 42} = 29
4	Time for Node 5 - Time for Activity G = 29 - 7 = 22

3	Min {Time for Node 5 - Time for Activity F, Time for Node 6 - Time for Activity E} = Min {29 - 8, 41 - 14} = Min {21, 27} = 21
2	Min {Time for Node 3 - Time for Activity B, Time for Node 4 - Time for Activity D, = Min {21 - 13, 22 - 12, 29 - 9} = Min {8, 10, 20} = 8
1	Time for Node 2 - Time for Activity A = 8 - 8 = 0

Latest Completion Times of the activities

Activity	Latest Finish Time (Weeks)
K	58
J	51
I	51
H	41
G	29
F	29
E	41
D	22
C	29
B	21
A	8

Calculation of Total Float for every activity:

Activity	Duration (Weeks)	Earliest Start Time	Earliest Finish Time	Latest Start Time	Latest Finish Time	Total Float = Latest Finish Time - Earliest Finish Time
A	8	0	8	0	8	0
B	13	8	21	8	21	0
C	9	8	17	20	29	12
D	12	8	20	10	22	2
E	14	21	35	27	41	6
F	8	21	29	21	29	0
G	7	20	27	22	29	2
H	12	29	41	29	41	0
I	9	29	38	42	51	13
J	10	41	51	41	51	0
K	7	51	58	51	58	0

The tasks with total float = 0 are A, B, F, H, J, and K. These are the essential activities.

The time required to complete the project = 58 weeks.

Calculation of the slacks of the events

Slack of an event = Latest Allowable Time of Occurrence of the event - Earliest

Expected Time of Occurrence of that event.

Event (Node)	Earliest Expected Time of Occurrence of Event	Latest Allowable Time of Occurrence of Event	Slack of the Event
1	0	0	0
2	8	8	0
3	21	21	0
4	20	22	2
5	29	29	0
6	41	41	0
7	51	51	0
8	58	58	0

Interpretation

Based on the constraints of the events, it is determined that the timing of event 4 might be postponed by a maximum of 2 weeks, while no other event can be deferred.

Questions

1. Describe the concepts: The minimum and maximum times for the activities of a project.
2. Clarify the method to calculate the earliest anticipated time of an event.
3. Clarify the method to calculate the latest permissible time of an event.
4. What does the slack of an activity refer to? How will you identify it?
5. Examine the project with the subsequent details:

Activity	Duration (weeks)
1→2	1
2→3	3
2→4	7
3→4	5
3→5	8
4→5	4
5→6	1

Ascertain the earliest and the latest times for the activities. Compute the total float for every activity along with the slacks of the events.

Crashing of a Project

Lesson Outline

- The Concept Of Project Crashing
- The Standards For Selecting An Activity To Crash
- Quantitative Problems

Learning Objectives

After going through this lesson, you ought to be capable of

- Comprehend the idea of project crashing
- Select an activity for crashing
- Solve numerical problems

The Meaning of Crashing

The procedure of reducing the duration needed to finish a project is referred to as crashing, and it is typically accomplished by utilizing extra labor or equipment for one or several activities. Crashing entails increased expenses. A project manager aims to accelerate a project while incurring as little additional cost as feasible. Project crashing strives to lessen the extra expenses for completing a project ahead of the designated deadline.

Steps in Project Crashing

Assumption: It is presumed that a linear relationship exists between time and cost.

Let us examine project crashing using the critical path method. The subsequent four-step procedure is utilized.

Step 1: Determine the critical path using the normal durations and standard expenses for the tasks and pinpoint the critical tasks.

Step 2: Determine the crash cost per unit time for each activity within the network. This can be calculated using the following formula.

$$\text{Crash Cost / Time period} = \frac{\text{Crash Cost-normal Cost}}{\text{Normal Time-crash Time}}$$

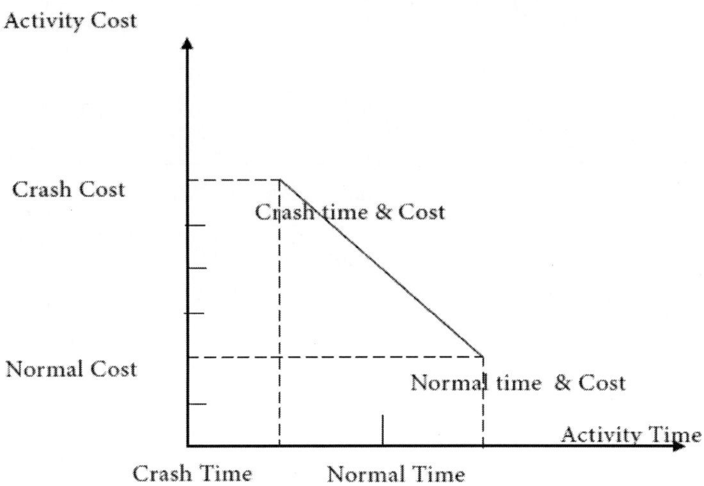

Step 3: Choose an activity for crashing. The criteria for the selection are as follows:

Choose the task on the critical path that has the lowest cost to crash per unit of time. Reduce this task to the greatest amount of time that is allowed by the provided information.

Crashing an activity necessitates additional expenses. Nevertheless, even if the organization is willing to invest more funds, the activity duration cannot be shortened past a specific threshold due to various other considerations. In step 1, it is important to recognize that decreasing the duration of an activity on the critical path is the only way to minimize the overall project completion time. For this reason, we choose an activity on the critical path for crashing. In step 3, we must contemplate the following question:

If we aim to decrease the project completion time by one unit, which essential activity will require the least extra cost?

Based on the minimum extra expense, a crucial task is selected for crashing. If there is a draw between two critical tasks, the tie may be settled in any manner.

Step 4: Following the crashing of an activity, identify which path is critical under the new conditions. At times, shortening the duration of an activity on the critical path can lead to a non-critical path becoming critical. If the initial critical path remains the longest path, then proceed to Step 3. If not, establish the new critical path and then move on to Step 3.

Example 36

A project includes tasks with the subsequent normal and crash durations and expenses:

Activity	Predecessor Activity	Normal Time (Weeks)	Crash Time (Weeks)	Normal Cost (Rs.)	Crash Cost (Rs.)
A	-	4	3	8,000	9,000
B	A	5	3	16,000	20,000
C	A	4	3	12,000	13,000
D	B	6	5	34,000	35,000
E	C	6	4	42,000	44,000
F	D	5	4	16,000	16,500
G	E	7	4	66,000	72,000
H	G	4	3	2,000	5,000

Establish a crashing scheme for the aforementioned project to decrease the overall project duration by 3 weeks.

Solution

We possess the subsequent network diagram for the specified project with standard costs:

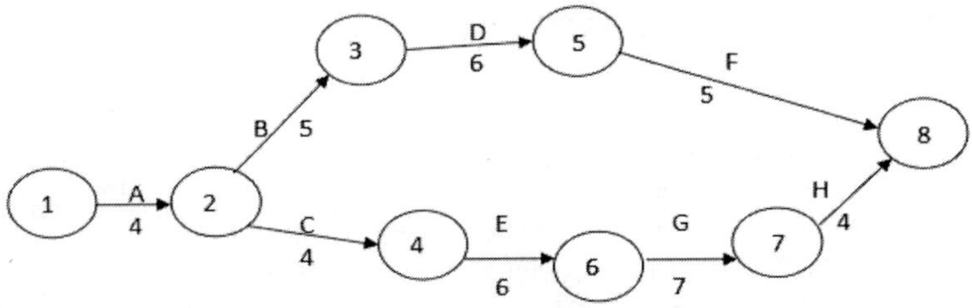

Starting at the Start Node and concluding at the End Node, there are two routes for the network as outlined below:

Path I

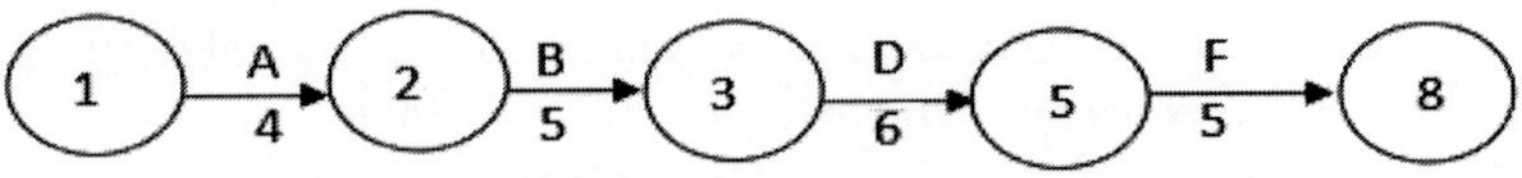

The duration for the route = 4 + 5 + 6 + 5 = 20 weeks.

Path II

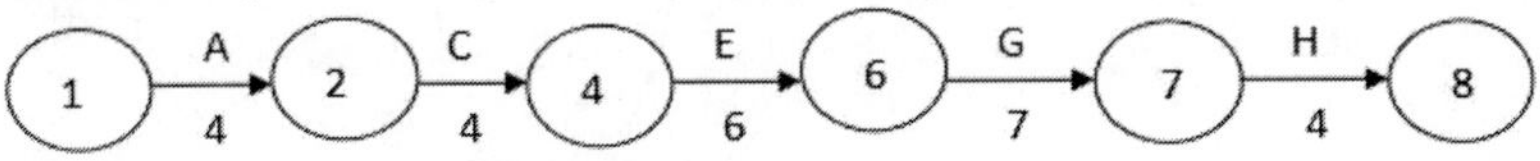

The duration for the route = 4 + 4 + 6 + 7 + 4 = 25 weeks.

Maximum of {20, 25} = 25.

Thus, Path II represents the critical path, and the essential activities include A, C, E, G, and H. The activities that are not critical are B, D, and F.

Given that the standard duration of activity A is 4 weeks while its expedited duration is 3 weeks. Therefore, the duration of this activity can be shortened by one week if the management is willing to invest an additional sum. Nevertheless, the time cannot be decreased by more than one week even if the management is willing to spend more funds. The standard cost of this activity is Rs. 8,000 while the expedited cost is Rs. 9,000. From this, we observe that reducing the duration of activity A by one week will incur an additional expense of Rs. 1,000 for the management. In a comparable manner, we can calculate the crash cost per unit time for the other activities as well. The findings are presented in the table below.

Activity	Normal Time	Crash Time	Normal Cost	Crash Cost	Crash cost - Normal Cost	Normal Time - Crash Time	Crash Cost per unit time
A	4	3	8,000	9,000	1,000	1	1,000
B	5	3	16,000	20,000	4,000	2	2,000
C	4	3	12,000	13,000	1,000	1	1,000
D	6	5	34,000	35,000	1,000	1	1,000
E	6	4	42,000	44,000	2,000	2	1,000

F	5	4	16,000	16,500	500	1	500
G	7	4	66,000	72,000	6,000	1	6,000
H	4	3	2,000	5,000	3,000	1	3,000

A non-essential task can be postponed without affecting the timeline of the entire project. However, if an essential task is postponed, it will hold up the entire project. Due to this reason, we need to identify a crucial activity for crashing. In this case, we must select one of the tasks A, C, E, G, or H. The cost for crashing per unit of time is calculated as follows:

Rs. 1,000 for A; Rs. 1,000 for C; Rs. 1,000 for E; Rs. 6,000 for G; Rs. 3,000 for H.

The lowest among them is Rs. 1,000. Therefore, we need to select an activity with Rs. 1,000 as the crash cost for each unit of time. Howeve', A, C, and E are tied. This tie can be resolved in any way. Let us choose A for the crashing process. We will decrease A's time by one week by investing an additional amount of Rs. 1,000.

Following this step, we possess the subsequent network with the updated durations for the tasks:

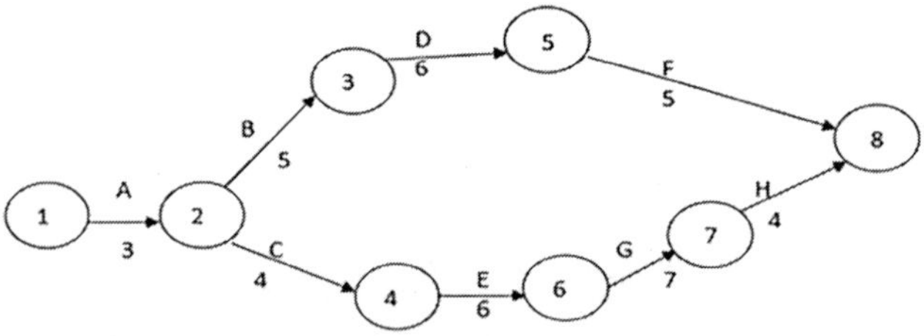

The updated duration for Path I = 3 + 5 + 6 + 5 = 19 weeks.

The duration for Path II = 3 + 4 + 6 + 7 + 4 = 24 weeks.

The greatest of {19, 24} = 24.

Therefore, Path II is identified as the critical path, and the critical activities include A, C, E, G, and H. Nonetheless, the duration for A cannot be decreased any further. As a result, we need to evaluate C, E, G, and H for potential crashing. Of these activities, C and E present the lowest crash cost per unit of time. The tie between C and E can be arbitrarily settled. Let's assume we decrease the duration of C by one week at an additional cost of Rs. 1,000.

After completing this step, we now possess the subsequent network with the updated durations for the activities:

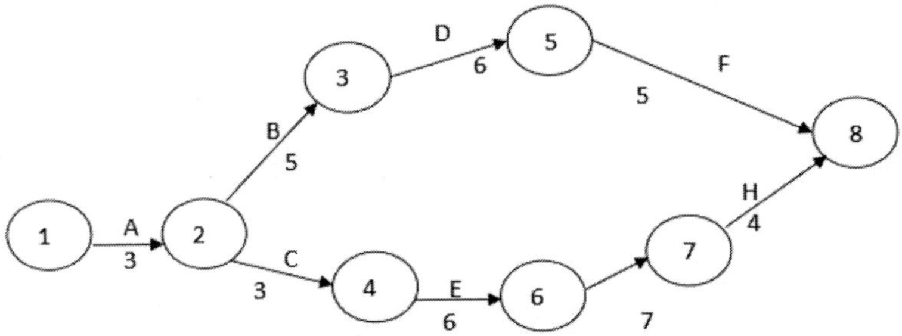

The duration for Path I = 3 + 5 + 6 + 5 = 19 weeks.

The duration for Path II = 3 + 3 + 6 + 7 + 4 = 23 weeks.

The greatest of {19, 23} = 23.

Therefore, Path II is the essential path and the vital activities are A, C, E, G, and H. Now the duration for A or C cannot be shortened any further. Thus, we need to look at E, G, and H for crashing. Of these, E has the lowest crash cost per unit of time. Consequently, we decrease the duration of E by one week at an additional cost of Rs. 1,000.

By the specified condition, we must decrease the project duration by 3 weeks. As this has been achieved, we conclude this phase.

Result: We have reached the subsequent crashing plan for the specified project:

Shorten the durations of A, C, and E by one week for each.

The project duration after crashing is 22 weeks.

Additional funds needed = 1,000 + 1,000 + 1,000 = Rs. 3,000.

Example 37

The leadership of a company is keen on accelerating the completion of the following project by investing an extra sum that does not surpass Rs. 2,000. Propose methods by which this can be achieved.

Activity	Predecessor Activity	Normal Time (Weeks)	Crash Time (Weeks)	Normal Cost (Rs.)	Crash Cost (Rs.)
A	-	7	6	15,000	18,000
B	A	12	9	11,000	14,000
C	A	22	21	18,500	19,000
D	B	11	10	8,000	9,000
E	C, D	6	5	4,000	4,500

Solution

We possess the subsequent network diagram for the specified project with standard costs:There are two routes for this project as outlined below:

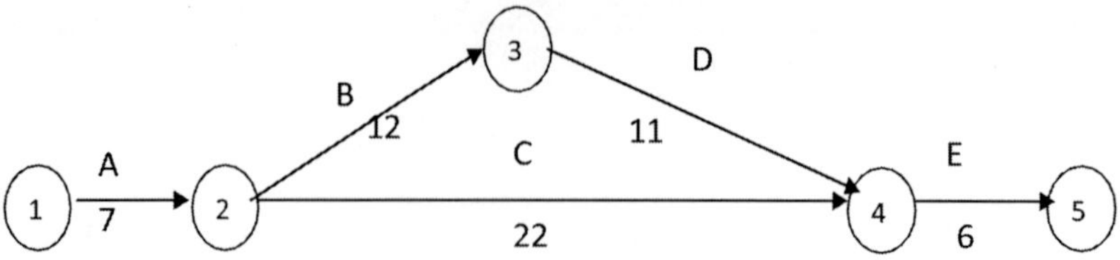

Path I

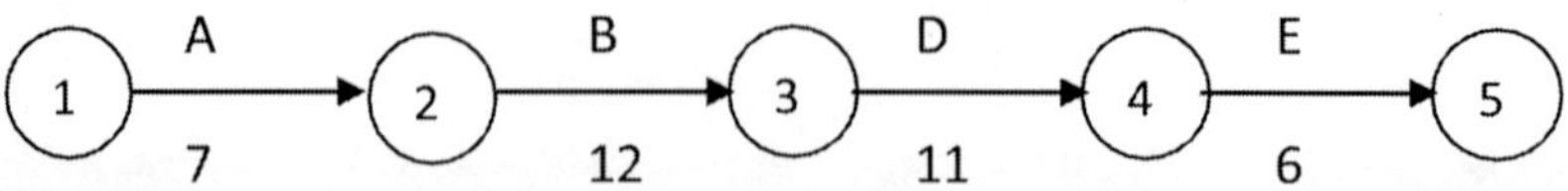

The duration for the route = 7 + 12 + 11 + 6 = 36 weeks.

Path II

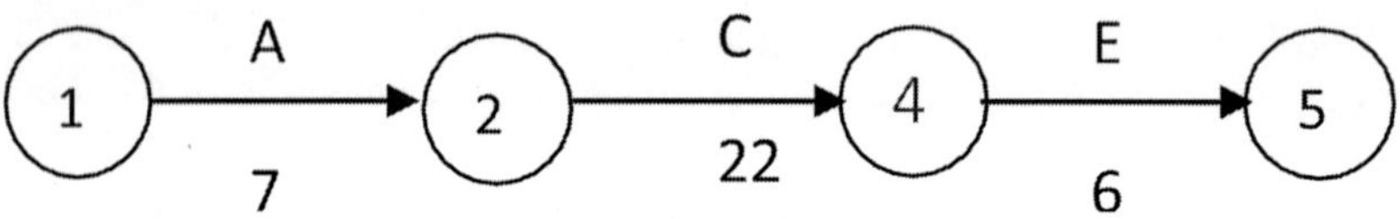

The duration for the route = 7 + 22 + 6 = 35 weeks.

Maximum of {36, 35} = 36.

Thus, Path I represents the critical path, and the critical activities include A, B, D, and E. The activity that is non-critical is C.

The unit time crash cost for the activities in the project is presented in the table below.

Activity	Normal	Crash Time	Normal Cost	Crash Cost	Crash cost - Normal Cost	Normal Time - Crash Time	Crash Cost per unit time
A	7	6	15,000	18,000	3,000	1	3,000
B	12	9	11,000	14,000	3,000	3	1,000
C	22	21	18,500	19,000	500	1	500
D	11	10	8,000	9,000	1,000	1	1,000
E	6	5	4,000	4,500	500	1	500

We need to select one of the options A, B, D, and E for crashing. The cost to crash per unit of time is detailed as follows:

Rs. 3,000 for A; Rs. 1,000 for B; Rs. 1,000 for D; Rs. 500 for E.

The minimum amount among them is Rs. 500. Therefore, we need to select activity E for crashing. We decrease the duration of E by one week by investing an additional Rs. 500.

After completing this phase, we possess the subsequent network with the updated durations for the tasks:

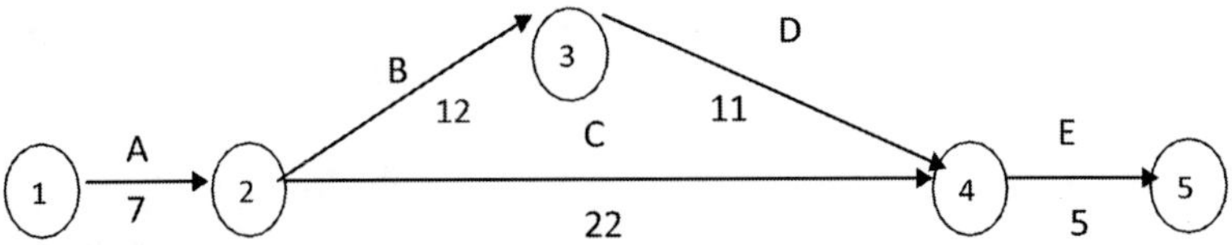

The updated duration for Path I = 7 + 12 + 11 + 5 = 35 weeks.

The duration for Path II = 7 + 22 + 5 = 34 weeks.

Highest of {35, 34} = 35.

Thus, Path I is the essential path and the vital activities are A, B, D, and E. The activity that is not critical is C.

The duration of E cannot be shortened any further. Therefore, we cannot choose it for crashing. Next, B and have the lowest crash cost per unit time. Let's choose B for crashing. Let's decrease the time of B by one week at an additional cost of Rs. 1,000.

After completing this step, we possess the subsequent network with the updated durations for the tasks:

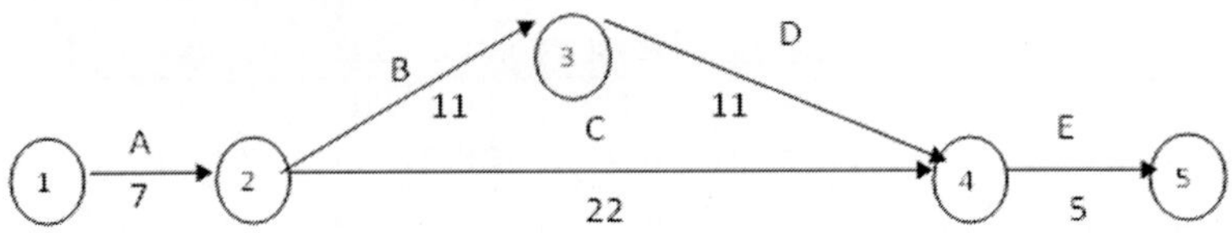

The updated duration for Path I = 7 + 11 + 11 + 5 = 34 weeks.

The duration for Path II = 7 + 22 + 5 = 34 weeks.

The largest of {34, 34} = 34.

Since both routes take the same amount of time, both are considered critical paths. Therefore, we can select an activity to crash from either path based on the lowest crash cost per unit of time. In path I, the activities include A, B, D, and E. In path II, the activities consist of A, C, and E.

The expense for crashing per unit time is the lowest for activity C. Therefore, we choose C for the crashing process. Decrease the duration of C by one week for an additional cost of Rs. 500.

By the specified condition, the additional amount must not go beyond Rs. 2,000. As this situation has been fulfilled, we conclude this step.

Result: The subsequent crashing plan is proposed for the specified project:

Decrease the duration of E, B, and C by one week each.

The project duration after crashing is 33 weeks.

Additional amount needed = 500 + 1,000 + 500 = Rs. 2,000.

Example 38

The manager of a firm intends to implement crashing for the upcoming project by allocating an extra budget that does not surpass Rs. 2,000. Provide your recommendation to the manager.

Activity	Predecessor Activity	Normal Time (Weeks)	Crash Time (Weeks)	Normal Cost (Rs.)	Crash Cost (Rs.)
A	-	20	19	8,000	10,000
B	-	15	14	16,000	19,000
C	A	22	20	13,000	14,000
D	A	17	15	7,500	9,000
E	B	19	18	4,000	5,000
F	C	28	27	3,000	4,000
G	D, E	25	24	12,000	13,000

Normal Distribution Table

Area Under Standard Normal Distribution

	0.00	0.01	0.02	0.03	0.04	0.05	0.06	0.07	0.08	0.09
0.0	0.0000	0.0040	0.0080	0.0120	0.0160	0.0199	0.0239	0.0279	0.0319	0.0359
0.1	0.0398	0.0438	0.0478	0.0517	0.0557	0.0596	0.0636	0.0675	0.0714	0.0753
0.2	0.0793	0.0832	0.0871	0.0910	0.0948	0.0987	0.1026	0.1064	0.1103	0.1141
0.3	0.1179	0.1217	0.1255	0.1293	0.1331	0.1368	0.1406	0.1443	0.1480	0.1517
0.4	0.1554	0.1591	0.1628	0.1664	0.1700	0.1736	0.1772	0.1808	0.1844	0.1879
0.5	0.1915	0.1950	0.1985	0.2019	0.2054	0.2088	0.2123	0.2157	0.2190	0.2224
0.6	0.2257	0.2291	0.2324	0.2357	0.2389	0.2422	0.2454	0.2486	0.2517	0.2549
0.7	0.2580	0.2611	0.2642	0.2673	0.2704	0.2734	0.2764	0.2794	0.2823	0.2852
0.8	0.2881	0.2910	0.2939	0.2967	0.2995	0.3023	0.3051	0.3078	0.3106	0.3133
0.9	0.3159	0.3186	0.3212	0.3238	0.3264	0.3289	0.3315	0.3340	0.3365	0.3389
1.0	0.3413	0.3438	0.3461	0.3485	0.3508	0.3531	0.3554	0.3577	0.3599	0.3621
1.1	0.3643	0.3665	0.3686	0.3708	0.3729	0.3749	0.3770	0.3790	0.3810	0.3830
1.2	0.3849	0.3869	0.3888	0.3907	0.3925	0.3944	0.3962	0.3980	0.3997	0.4015
1.3	0.4032	0.4049	0.4066	0.4082	0.4099	0.4115	0.4131	0.4147	0.4162	0.4177
1.4	0.4192	0.4207	0.4222	0.4236	0.4251	0.4265	0.4279	0.4292	0.4306	0.4319
1.5	0.4332	0.4345	0.4357	0.4370	0.4382	0.4394	0.4406	0.4418	0.4429	0.4441
1.6	0.4452	0.4463	0.4474	0.4484	0.4495	0.4505	0.4515	0.4525	0.4535	0.4545
1.7	0.4554	0.4564	0.4573	0.4582	0.4591	0.4599	0.4608	0.4616	0.4625	0.4633
1.8	0.4641	0.4649	0.4656	0.4664	0.4671	0.4678	0.4686	0.4693	0.4699	0.4706
1.9	0.4713	0.4719	0.4726	0.4732	0.4738	0.4744	0.4750	0.4756	0.4761	0.4767
2.0	0.4772	0.4778	0.4783	0.4788	0.4793	0.4798	0.4803	0.4808	0.4812	0.4817
2.1	0.4821	0.4826	0.4830	0.4834	0.4838	0.4842	0.4846	0.4850	0.4854	0.4857
2.2	0.4861	0.4864	0.4868	0.4871	0.4875	0.4878	0.4881	0.4884	0.4887	0.4890
2.3	0.4893	0.4896	0.4898	0.4901	0.4904	0.4906	0.4909	0.4911	0.4913	0.4916
2.4	0.4918	0.4920	0.4922	0.4925	0.4927	0.4929	0.4931	0.4932	0.4934	0.4936
2.5	0.4938	0.4940	0.4941	0.4943	0.4945	0.4946	0.4948	0.4949	0.4951	0.4952
2.6	0.4953	0.4955	0.4956	0.4957	0.4959	0.4960	0.4961	0.4962	0.4963	0.4964
2.7	0.4965	0.4966	0.4967	0.4968	0.4969	0.4970	0.4971	0.4972	0.4973	0.4974

2.8	0.4974	0.4975	0.4976	0.4977	0.4977	0.4978	0.4979	0.4979	0.4980	0.4981
2.9	0.4981	0.4982	0.4982	0.4983	0.4984	0.4984	0.4985	0.4985	0.4986	0.4986
3.0	0.4987	0.4987	0.4987	0.4988	0.4988	0.4989	0.4989	0.4989	0.4990	0.4990

4

Financial Appraisal

Investment Appraisal Techniques

A financial evaluation/ appraisal of Project considers the possible compensations of completing the Project against the anticipated expenses. The assessment of the project will rely on several factors, including the Size of the project, time span of project etc. The motivation behind the financial evaluation is to decide if the project is advantageous, contrasting its expenses and its expected normal advantages. These methods are primarily designed to evaluate the effectiveness of a new project. This investment evaluation technique includes Payback Period, Internal Rate of Return (IRR), Net Present Value (NPV), Accounting Rate of return (ARR), Profitability index etc. Each of these techniques evaluates the project from various angles and provides very useful insights. Investment appraisal techniques are divided in two parts namely, discounted cash flow methods and non-discounted cash flow methods.

4.1 Non- Discounted Cash Flow Techniques

Future cash flows are not brought to their Present value through Non-Discounted Cash flow methods. Non-Discounted Cash flow methods are also categorized into the following two techniques.

a) Payback Period
b) Accounting Rate of Return
c) Ranking by Inspection

A. Payback Period

Payback period is one of the simplest investment appraisal techniques. The payback technique states what amount of time it requires for the undertaking to produce adequate income to take care of the underlying expense of the Project. Another simple method to assess a project is by evaluating the time required to recoup the investment made in the project. The payback period for the project can be calculated using the formula below:

$$P = \frac{I}{E}$$

Where,

P = the reimbursement duration of the initiative in years,

I = Project investment in Rs.

E = annual net cash flow at Rs.

The concrete project takes precedence due to the quicker payback period. This is demonstrated in table 4.1

Table 4.1: Estimation of payback Period Initial investment = Rs. 20,000

Year	Cash Flow (in Rs.)	
	Project 'A'	**Project 'B'**
0	-20,000	-20,000
1	5,000	4,000
2	5,000	4,000
3	5,000	4,000
4	5,000	4,000
5	5,000	4,000
6	5,000	4,000

$$\text{Project 'A'} = \frac{\text{Rs.}20{,}000}{\text{Rs.}5{,}000} = 4\,\text{years}$$

$$\text{Project 'B'} = \frac{\text{Rs.}20{,}000}{\text{Rs.}4{,}000} = 5\,\text{years}$$

Selecting from available options is inadequate since it overlooks the crucial factors like job security, time of income, return after time price, and if upcoming cash flows are either positive or negative.

Example 1

Pal Enterprise is planning to purchase machine of Rs.100000. There are two devices accessible for the same work and same cost, Machine X and Machine Y. The revenue data of two machines are given below.

Year	Revenue generated by Machine X	Revenue generated by Machine Y
1	Rs. 50000	Rs. 30000
2	Rs. 50000	Rs. 40000
3	Rs. 20000	Rs. 30000

Solution

As per the above data, The Payback Period of Machine X is 2 years as the machine cost Rs.100000 recovers in 2 Years (Revenue Rs 50000 for year 1 and Revenue Rs 50000 for year 2)

The Payback Period of Machine Y is 3 years as the machine cost Rs.100000 recovers in 3 Years (Revenue Rs 30000 for year 1, Revenue Rs 40000 for year 2 and Revenue Rs.30000 for year3)

According to Payback Period, Machine X will be given preference to purchase.

B. Accounting Rate of Return (ARR)

Accounting rate of return method is utilized to determine anticipated profit from an investment. It represents the net accounting profit generated from the investment as a proportion of that capital investment. It is also referred to as return on investment or return on capital.

The accounting rate of return is determined using the following formula

Accounting Rate of Return (ARR) = (Average annual profit after tax / Initial investment) x 100

Example 2

PAL Enterprise is hoping to put resources into a Machine. The new machine, which costs Rs. 420,000, would increment yearly income by Rs. 200,000 and yearly cost by Rs. 50,000. The machine is assessed to have a helpful existence of 12 years.

Solution

Deterioration cost each year (Depreciation Cost per Year) = Rs. 420,000/12 = Rs. 35,000

Expansion in average yearly benefit (Average annual Profit) = Rs. 200,000 - (Rs. 50,000 + Rs. 35,000) = Rs. 115,000

Initial investment = Rs. 420,000

ARR = Average annual Profit / Initial Investment

= (Rs. 115,000/ Rs. 420,000) * 100

= 27.38%

There is an opportunity for PAL Enterprise to have 27.38 % expected profit from an Investment in a Machine.

C. Ranking by Inspection

This is contingent upon the magnitude of the expenses and the length of the loan. If two initiatives share the same investment and the same net output value but vary in duration, the project with the longer duration is preferable to the one with the shorter duration. This understandably results in a skewed selection arising from an absence of thorough and precise analysis.

3.2 Discounted Cash Flow Techniques

Discounted cash flow methods assess the current worth of future cash flows that will be received. Cash flows represent the yearly profits produced by the project. When evaluated at a discount rate, it constitutes a discounted cash flow. These discounted cash flows provide the most accurate estimation for assessing the project's value. This method yields the current worth of the project. The present value of the expenditures is deducted from the present value of the advantages to calculate the present value of the plan for each year.

The following are classified under discounted cash flow techniques:

a) Net present value
b) Internal rate of return
c) Profitability index
d) Discounted payback period

A. Net Present Value (NPV)

It is the most widely regarded conventional method for investment assessment (Investment appraisal). Net present value refers to the quantity of discounted future cash inflows and outflows associated with the Project. By and large, the Weighted Average Cost of Capital (WACC) is the discounting component for future incomes in the net present value method. It refers to the standard present value of the cash flow. Often referred to as net present value (NPV). The criterion for selecting a project relies on the positive NPW value when it is discounted by the opportunity cost of capital. It can be executed successfully if there is a precise estimation of the opportunity cost of capital. NPW serves as a definitive measure, however, it is not a comparative one.

This method sums up the discounted net incomes from the investment by the lowest necessary rate of return and subtracts the initial investment to provide the 'net present value'. The organization ought to acknowledge the task assuming that the NPV is positive. NPV should be Positive and Company or managers should work on the project if the NPV is Positive.

$$\textbf{NPV} = \sum \frac{\text{Cash Flow}_t}{(1+i)^t} - \text{Initial Investment}$$

Where:

- Cash Flow_t = Cash flow at time period t
- i = Discount rate
- t = Time period
- Initial Investment = Initial cost of the project

Example 3

PAL Enterprise is begins the project at a cost of Rs. 100000. The Project will generate the income of Rs. 40000 for Year 1, Rs. 50000 for Year 2 and Rs. 50000 for Year 3. The PAL enterprise's WACC is 10%. Figure out NPV.

Solution

NPV = Cash flow / ((1+i) ^ t – Initial Investment

NPV = {Rs.40000 / $(1 + 0.1)^1$} + {Rs.50000 / $(1 + 0.1)^2$} + {Rs.50000 / $(1 + 0.1)^3$} – Rs. 100000

= {Rs. 36363.63 + Rs. 41322.31 + Rs. 37565.74} - Rs. 100000

= Rs. 115251.68 – Rs.100000

= Rs. 15251.68

In this case, the Net present value of a Project is Rs. 15251.68 and it is Positive therefore the PAL enterprise should accept the project to work.

B. Internal Rate of Return (IRR)

Internal Rate of Return (IRR) equates the discounted future cash flow with the original investment. It is the discount rate at which the company experiences neither a profit nor a loss. It will obtain by the trial and error techniques. It can also state that IRR is the rate at which the NPV of a Project equals zero. This indicates that the Present value of Cash inflow minus the Present value of cash outflow is equal to Zero.

When calculating the internal rate of return (IRR), the time value of money is calculated. The IRR method provides knowledge about the return on various projects. So IRR is also referred to as "marginal return" of capital or return on investment. This represents the discount rate at which the present values of the net cash flows are equal to zero, i. e. NPW = zero. This IRR must be found by trial and error with some approximations. This procedure for the cereal and mango programs is explained in Tables 4.2 and 4.3.

In the procedure, a discount rate is taken into account and the related NPW is calculated. A favorable NPW figure for a project suggests that the IRR remains elevated and the subsequent necessary IRR figure needs to exceed the initial level. This approach proceeds until the NPW is compromised. Afterwards, by employing the integration method, the accurate IRR is determined utilizing the subsequent equation.

$$[\text{Internal rate of return}] = [\text{Lower discount rate}] + [\text{Difference between the two discount rates}] \times \left[\frac{\text{Present worth of the cash flow at the lower discount rate}}{\text{Absolute difference between the present worths of the cash flow at the two discount rates}}\right]$$

Table 4.2: Estimation of IRR for Sericulture (One Hectare) (Hypothetical)

Year	Costs (in Rs.)	Gross income (in Rs.)	Net income (in Rs.)	Discount factor (40%)	Net present worth (in Rs.)	Discount factor (43%)	Net present worth (in Rs.)
1	38,900	--	-38,900	0.7143	-27,786.27	0.6993	-27,202.77
2	9,239	28,475	19,236	0.5102	9,814.21	0.48902	9,406.4
3	10,575	32,550	21,975	0.3644	8,007.69	0.3419	7,513.25
4	11,952	35,610	23,658	0.2603	6,158.17	0.2391	5,656.62
5	12,858	39,802	26,944	0.1859	5,008.89	0.1672	4,505.04
			52,913		1,202.69		-121.46

NB: The life expectancy of the mango plantation needs to be taken into account in order to work with the IRR. Due to lack of information, we only consider seven years here for photo purposes.

$$\text{IRR} = 40 + 3\left[\frac{1202.69}{1202.69 + 121.46}\right]$$

= 40 + 3(0.9083)

= 40 + 2.7249

= 42.7249%

= 42.7%

Table 4.3: Estimation of IRR for Mango Orchard (One Hectare) (Hypothetical).

Year (in Rs.)	Costs (in Rs.)	Gross income (in Rs.)	Net income (in Rs.)	Discount factor (25%)	Net present worth (in Rs.)	Discount factor (30%)	Net present worth (in Rs.)
End of 6th year	25,000	--	-25,000	0.262	-6,550	0.207	-5,175.00
End of 7th year	4,250	10,260	6,010	0.21	1,262.01	0.159	955.59
End of 8th year	4,792	12,550	7,758	0.168	1,303.30	0.123	954.23
End of 9th year	5,368	14,530	9,162	0.134	1,227.71	0.094	861.23
End of 10th year	5,975	16,275	10,300	0.107	1,102.10	0.073	751.90
End of 11th year	6,456	19,396	12,940	0.086	1,112.84	0.056	724.64
End of 12th year	7,187	21,470	14,283	0.069	985.53	0.043	614.17
			35,453		443.49		-313.24

$$\text{IRR} = 25 + 5\left[\frac{443.49}{443.49 + 313.24}\right]$$

= 25 + 5(0.586)

= 25 + 2.93

= 27.93%

C. Profitability Index (PI)

Profitability Index is among the straightforward investment appraisal techniques; it helps us to identify how much we can earn per Rupees of Investment. Profitability Index can be calculated as division of the Current Worth of a projected future cash inflow and initial investment of the Project.

Profitability Index = Present Worth of a projected future cash inflow of the Project / initial expenditure of the Project

The Profitability Index of a Project is Greater than 1, the Project should be accepted and if it is less than 1 the Project should be avoided.

Example 4

Assume any Project that have, the current worth of expected future cash flow/income is Rs. 120000 and the initial outflow is Rs. 100,000. Find out Profitability Index what it implies?

The Profitability Index of a Project is Greater than 1, the Project should be accepted Profitability Index = Present Value of an anticipated future cash flow / initial outflow

= Rs. 120000 / Rs. 100,000

= 1.2

It implies that each invested Rupees is creating an income of 1.2 rupees and the Profitability Index of a Project is Greater than 1 therefore the Project should be accepted.

D. Discounted Payback Period Method

This technique is equivalent to the payback period method. The main difference in both methods is, in the Discounted Payback Period Method the payback period is determined based on discounted future cash-flows while in Payback Method the payback period is calculated based on anticipated cash flows.

Benefit-Cost Ratio

Here, the current value of the costs is contrasted with the current value of the benefits. The real value of the profit-to-cost ratio varies according to the interest rate stated. Even though the project choice is determined by the B-C ratio, the primary approach for project selection is to choose a project that has a B-C ratio exceeding one, which reduces the opportunity cost of capital. Ultimately, the chosen project is picked for execution among options based on the highest B-C ratio.

The benefit-cost ratio is a measurement derived by dividing the total of discounted benefits by discounted costs.

Steps for determining the benefit-cost ratio

Step 1: Calculate the cash inflows and outflows on an annual basis.

Step 2: Determine the discount value of 1 for specific years at the provided discount rate.

Step 3: Multiply the cash inflows and cash outflows for each year by the relevant discount factor.

Step 4: Calculate the total of the discounted amounts for cash inflows and outflows individually.

Step 5: Divide the reduced values of cash inflows by cash outflows to achieve the benefit-cost ratio.

$$\text{B-C Ratio} = \frac{\sum_{t=1}^{n} \frac{B_t}{(1+r)^n}}{\sum_{t=1}^{n} \frac{C_t}{(1+r)^n}}$$

The following example shows an estimate of the B-C ratio.

Table 4.4: Estimation of Benefit-cost Ratio (BCR) for Sericulture (Hypothetical) (one ha)

Year	Costs (in Rs.)	Gross Returns (in Rs.)	Discount Factor at 12%	Present worth of costs (in Rs.)	Present worth of gross returns (in Rs.)
1	38,900	-	0.8929	34,733.81	-
2	9,239	28,475	0.7972	7,365.33	22,700.27
3	10,575	32,550	0.7118	7,527.29	23,169.09
4	11,952	35,610	0.6355	7,595.50	22,630.16
5	12,858	39,802	0.5674	7,295.63	22,583.65
				64,517.56	91,083.17

$$\text{Benefit} - \text{cost} - \text{ratio} = \frac{\text{Present worth of gross returns}}{\text{Present worth of costs}} = \frac{91083.18}{64517.56} = 1.41$$

Table 4.5: Estimation of Benefit-cost Ratio (BCR) for mango orchard (Hypothetical) (one ha)

Year	Costs (in Rs.)	Returns (in Rs.)	Discount Factor at 12%	Present worth of costs (in Rs.)	Present worth of gross returns (in Rs.)
At the end of 6th year	25,000	-	0.507	12,675.00	--
At the end of 7th year	4,250	10,260	0.452	1,921.00	4.637.52
At the end of 8th year	4,792	12,550	0.404	1,935.97	5,070.20
At the end of 9th year	5,368	14,530	0.361	1,937.85	5,245.33
At the end of 10th year	5,975	16,275	0.322	1,923.95	5,240.55
At the end of 11th year	6,456	19,396	0.287	1,852.87	5,566.55
At the end of 12th year	7,187	21,470	0.257	1,847.06	5,517.79
				24,093.70	31,278.04

$$\text{Benefit} - \text{cost} - \text{ratio} = \frac{\text{Present worth of gross returns}}{\text{Present worth of costs}} = \frac{31278.04}{24093.70} = 1.30$$

5

Project Implementation and Control

Project Implementation

Project Implementation is the process of putting a project plan into action. It takes place after the planning phase; planning phase includes key objectives, timeline and budget of the project. Execution includes organizing resources and assessing performance to guarantee the project stays within its anticipated scope and budget. It also involves handling any unforeseen issues in a way that keeps a project running smoothly.

Steps for Project Implementation

1. **Assess the Project Plan:** Before implementing the project, it is essential to create a plan that satisfies the expectations of the management, customers, and other stakeholders. The project manager must assess the project plan and guarantee that all team members comprehend the project and its requirements. The project manager must define the roles and the time required to complete each role. This step sets the standard for cooperation at work.
2. **Execute the Plan:** At this stage, the real work on the project begins. The project manager regularly discusses progress with his team. It measures progress along a timeline and tracks resource usage.
3. **Make required Changes:** Sometimes, during the implementation of the project, the manager has to make changes at the request of the customers or in other situations. Make these changes as needed, based on the design plan for solutions. Continue to interact with the team and ask questions to determine areas where they need more support.
4. **Examine Project Data:** It is crucial to gather and evaluate data throughout the execution phase of the project to assess the advancement of the project in comparison to its initial plan. Now, the day manager can use many project management software and information technology to collect and analyze real-time data.
5. **Assess the Feedback:** After the conclusion of the project, it is crucial to gather feedback regarding the project outcomes from the clients, the project team and other stakeholders. This assists in determining the number of activities in the project that were executed as intended and aspects that require enhancement moving forward.
6. **Prepare final Project reports:** In this phase, project reports are created and shared with clients, stakeholders and the team, which describe how the various

activities performed against the planned budget and schedule. Indicates areas where further improvement is possible within scope, time and resources. This stage enables organizations to contemplate project accomplishments and recognize potential enhancements that could provide enduring advantages for the project management process.

Cost overruns

Cost overrun refers to an unanticipated change in the project that raises the overall project cost. These are the main causes of Overrun.

- Economic factors: errors in the scope or budget of the project.
- Technical factors, such as imprecise data gathering or estimations.
- Psychological factors: insufficient dedication to the project.

Primary Reasons or Caused for Cost Overrun in Projects

Followings are the primary reasons or caused for cost overrun in Projects

1. **Inaccurate project estimates:** Inaccurate project estimates may lead to both overestimation and underestimation. For projects, both are detrimental. Estimates need to be highly accurate and based on specific facts in order for the project to be completed properly.
2. **Scope creep:** Scope creep is common due to shifting business requirements and a dynamic business environment. A thorough and organised project tracking system is needed to combat scope creep. A methodical approach to project tracking can help manage scope creep in the project by keeping an eye on both finished and unfinished tasks.
3. **Lack of risk management:** Unpredictable internal and external factors can occasionally cause project delays or cost overruns. Consequently, implementing a risk and change management plan inside the project might help prevent unforeseen financial losses.
4. **Communication gaps within teams:** Ineffective communication between the project manager and stakeholders can result in schedule delays, misallocated resources, wasted effort, and project management cost overruns.

Project Controls

Project management is the process of collecting and analyzing project data to achieve estimated costs and schedules. Project management includes processes for measuring project status, predicting likely outcomes, and improving project performance. Project management tasks include initiating, planning, monitoring, controlling, communicating and closing project costs and schedules. The main purpose of project inspection is to minimize cost and plan deviations from the original plan. Project management includes the following tasks

- Connect projects to company/organizational goals and objectives
- Preparation of work breakdown structure (WBS)

- Work with the original project schedule
- Create a risk management strategy
- Evaluation of the project budget and forecasts
- Project cost monitoring
- Reporting and feedback
- Optimization of project strategies

Processes of Project Controls

A. Project Planning

Controllers and project managers work together in project planning. Planning includes preparation of project objectives, project plan, task plan, creation of work breakdown structures, assignment of roles and responsibilities, identification of stakeholders, cost estimates, etc. It offers an initial reference for engaging with all project execution tasks.

B. Preparation of the budget

Budgeting is very important to accurately calculate costs and understand when and why deviations occur.

C. Risk management

Project management offers a careful approach to risk management. By proactively identifying risks, continuously monitoring risks and developing contingency plans to resolve and mitigate issues, impact on budget and schedule can be minimized. It also helps prevent some risks that may arise in the future.

D. Change management

Change management is very important because a project deviates from its original estimates not because of a single factor, but instead because the variation is caused by the combined impact of multiple factors that are typically overlooked. It monitors changes and understands their impact by performing evaluation, approval and accountability, etc.

E. Prediction

Accurate forecasting can provide a wealth of information about drivers that lead to cost overruns and schedule delays. It additionally offers a standard for assessing the actual and pledged tasks of the project. Consistent and prompt project updates enable the project manager to respond swiftly and assist in implementing corrective measures if the project lags behind schedule.

F. Results management

It is very important to define and measure the key performance indicator that would guide the project.

G. The head of the project

Project management creates processes and systems that allow team members to communicate and collaborate with each other. The goal of project management is to track innovation, limit meeting minutes and learning, and manage appropriate workflows that allow team members to focus on real tasks instead of routine tasks.

Benefits of Project Controls

Here are some of the main benefits of project management:

- Reduction of project costs and duration
- On-time finishing of the project
- Enhance forecasting of project costs and duration
- Greater clarity regarding the project's financial status at every stage
- Ability to diminish the project's scope
- Improves project performance
- Increases job satisfaction of team members
- Reduce the level of risk

Project Controls and Information System

Project Management Information System (PMIS)

A project management information system gathers, arranges, and utilizes project information via one or multiple software applications. These applications assist project managers in planning, executing, and finalizing their projects while overseeing the flow of information. There exists a variety of PMIS software, the majority of which share similar functionalities like data collection, work authorization, scheduling, and delivery tools, among others. Some extra features include automated collection and reporting of key performance indicators (KPI).

What Is a PMIS Used For?

A project management information system is created to assist with every facet of project management, encompassing data that must be monitored or gathered. PMIS can serve for managing project scope, integration management, project time management, project cost management, project quality management, project risk management, communication management, and additional areas.

Essential Features of a PMIS

If it is necessary to organize and control an information tsunami for the seamless advancement of the project, the project management information system becomes a crucial resource for efficient project management. The data also assists in lowering expenses, enhancing productivity, and minimizing risks in all upcoming projects. The software includes numerous tools for data storage, facilitating straightforward access, analysis, search and thoroughly organize.

Schedule and Planning: This is one of the primary responsibilities of project management. PMIS determines the critical path, available time, and the early and late schedule.

Resource Management: Project resources include everything from materials to people. A project management information system supports this resource management involving resource entry, allocation, resource leveling, and more. Furthermore, it assists in monitoring the tasks and costs involved in procurement and distribution.

Budget: PMIS monitors all expenses linked to specific tasks, as the overall project and budget rely on the assessment of each task. To achieve a more precise budget calculation, the client's finances, allocated resources, and available time are considered, followed by the preparation of the budget.

Management and Performance: PMIS offers performance management through the examination and regulation of costs and performance, revising current plans when new data emerges or existing data alters, and supplying what-if scenarios to the project manager.

Reporting and Communication: The Project Management Information System evaluates data and generates reports for every task. It additionally creates charts and graphs from all the collected data and distributes the outcomes with the team and other pertinent individuals. It allows team members to interact efficiently with each other.

Integration and Usability: PMIS can even combine data with other systems, such payroll and inventories, and retrieve information from additional projects for cross-project evaluation. The cost and duration of training are directly correlated with a PMIS's ease of operation.

How Does PMIS in Project Management Work?

A PMIS carries out distinct tasks based on the project's stage. You should be able to monitor the duration, scope, and cost of your project with a fundamental PMIS. A good PMIS should incorporate various project management software functions that you can utilize at any stage of the project life cycle.

A. Project Initiation Phase

During] the initiation phase, a PMIS can assist in the following areas:

- Estimating an initial budget for the project that covers cost and resource assessments
- Defining the project scope and creating bids
- Timing project tasks and assigning them to the suitable team members
- Arranging project information and creating necessary reports for delivery to primary project stakeholders

B. Project Planning Phase

During the planning stage, a PMIS assists with the following:

- In-depth project scheduling, encompassing evaluation of critical paths and related tasks

- Funding and budget oversight support, involving the execution of budget analysis, cost management, and additional KPI measures
- Executing resource planning for the entire project, pinpointing current resources, and formulating a contingency plan for those that may be necessary in the future
- Establishing baseline metrics for the project's duration, expenses, and scope

C. Project Execution Stage

A PMIS is crucial for the effective implementation of projects. Furthermore, interacting with all stakeholders and team members, and monitoring discussions and replies is facilitated by its assistance.

- All team discussions, documents, and files are consolidated in a single hub that is accessible securely and easily by all stakeholders.
- This facilitates a straightforward comparison between the current project data and the baseline estimations of the project.
- Project managers have the ability to assess cost, budget, and schedule forecasts halfway through the project and adjust them according to the current project needs.
- It contributes to the efficient completion of various project modules, such as project reporting, cost management, material management, and project performance assessment.

D. Project Review and Closure Stage

PMIS aids in evaluating if the project aims and objectives were achieved during the project evaluation and completion phase.

- Facilitates a comprehensive assessment to ensure that the project's objectives were met or surpassed.
- All project details are systematically arranged and kept in a central repository for future retrieval and assessment.
- All project data is preserved for upcoming projects.
- Assists in generating productivity analysis measures and concluding project documents for stakeholder decision-making..

Advantages of Project Management Information System

There are two categories of PMIS: online or cloud-based and manual. The benefits of the former surpass all the advantages of a manual PMIS. You are able to oversee complex projects with the speed, efficiency, capacity, accuracy, and cost-effectiveness of an online PMIS—with speed being its most beneficial characteristic.

a) **Keeps the Project on Track:** You may steer clear of costly project delays by keeping track of budgets, prices, and schedules. Project oversight and prompt alerts assist in maintaining the project's timeline. The PMIS will immediately inform you via SMS or emails if there are any inconsistencies or if the project's progression appears to be on the verge of deviating from its intended path. The PMIS provides fast updates, allowing the team to quickly adopt new processes, reallocate resources, and make any necessary adjustments.

Teams can assess their performance after putting the necessary corrective measures in place. Businesses can gain a competitive edge by using the PMIS fast-tracking system to help them manage projects on time and under budget.

b) **Fosters Better Collaboration and Teamwork:** When team members are collaborating with different versions of the document, it can easily cause delays or derail the project. By sharing the most recent document around the team, PMIS prevents this problem. Documents can be controlled, shared, and updated by project managers, and the team will be notified when new versions are accessible.

c) **Gives a competitive advantage:** The project management information system is founded on data. Companies that are data-driven have a 23 times higher likelihood of gaining new customers, which translates to 19 times greater chances of obtaining above-average profits when compared to businesses that do not utilize data-driven decision-making. PMIS is an essential element in the success of project management that enhances organizations and capabilities in relation to their competitors. PMIS provides the team with access to all the information they may require to initiate tasks. Consequently, key decision-makers are able to access everything necessary, from intricate design processes to product testing outcomes and team hours. A comprehensive perspective of the project and its detailed data aids project managers in making improved decisions for the team. Along with providing the project with a noteworthy competitive edge, it can also enhance organizational efficiency and contribute to the accomplishment of project objectives.

d) **Manage Diverse Projects at the Same Time:** PMIS assists in the simultaneous management of several projects. Project managers possess a clear understanding of the advancements achieved in each project, as well as the schedule and resource allocation. Regardless of whether the project has been finalized, is currently ongoing, or is in the planning phase, stakeholders can monitor everything.

6

Introduction to AGRI Entrepreneurship

Entrepreneurship serves as a crucial element of economic development. In times of economic crisis, the significance of business development rises. Entrepreneurship has been linked to enhanced economic growth, greater wealth, and an elevated quality of life. In developing countries like India, the planning and implementation of entrepreneurship program development is essential for the standard of living of most of the backward areas due to their over-dependence on agriculture for employment (Uplaonkar and Biradar 2015). Entrepreneurship development therefore seems to be the best option to find job opportunities, generate income, reduce poverty and improve nutrition, health and overall food security in the national economy.

Entrepreneurship refers to the ability to take big risks to make a profit, to succeed, to manage and organize a new business. Entrepreneurship is a process in which a person becomes aware of owning a business as an alternative and develops new ideas for business. In addition, they try to learn the whole process of becoming an entrepreneur, and an initiative has been taken to develop entrepreneurship. It shows the practical application of all qualities such as innovation, creativity and risk in a work environment. The development of agriculture requires innovations in agricultural business. Stabilizing innovations and increasing productivity is a long-term and difficult business.

Business around the world is agriculture. Low productivity and latent unemployment are two inherent problems of Indian agriculture. However, the term "Agripreneur" is gaining popularity by leaps and bounds. The term "entrepreneur" is quite new in the field of agriculture. However, an agricultural entrepreneur is a person who manages and develops agricultural products.

Agriculture is considered the main economic activity that increases the general welfare of the country. In the past, agriculture was considered a low-tech industry dominated by many small family businesses that mostly focused on making things better rather than creating new things. However, in the last two decades, the situation has changed drastically due to the liberalization of the economy and the rapidly changing society. Agricultural enterprises must adapt to changing market demands, changing consumption patterns, food security sustainability and more. These changes opened the way for new entrants, innovations and portfolio entrepreneurship. Farmers, researchers, agribusinesses and governments have recognized this and emphasized a more entrepreneurial agricultural environment (De Lauwere et al_, 2002; McElwee, 2008; Pyysiainen *et al.* 2006). Agricultural entrepreneurship has a significant impact on the growth and survival of companies (Verhees *et al.*, 2011). Therefore, it invites both small and large farmers to engage in entrepreneurial agriculture.

Agribusiness is also known as Agricultural business. Agribusiness is generally defined as sustainable, community-based, direct market agriculture. It means a term that also refers to marketing and manufacturing of various agricultural products and production inputs. Small farmers cultivate agricultural goods for their households and additionally sell extra agricultural products to various markets where the efforts of small farmers demonstrate that they are agricultural entrepreneurs.

An important change has been noticed in the agricultural market and the exchange of agricultural products in various parts of the world in recent years. The world is evolving and moving from national markets to global exchanges. Farmers must develop farm plans together with other individual farmers and work as a joint stock company, not as a group. It helps to analyze different markets, sales, necessary financing and create different business opportunities with big customers. The concept of agribusiness is ultimately to support the farmers, the group of farmers and also the agricultural industry by improving production methods, improving the use of technology, providing environmental and market information.

The entrepreneurial skills of farmers must be developed and addressed by all stakeholders in the agricultural socio-economic network. Farmers have different strategies to survive and change the economic environment that lead to business growth. Agribusiness can be expanded, for example, through tourism or other non-agricultural business, or by integrating the value chain through food processing, direct marketing or organic production. The social and economic environment of agriculture should not be underestimated when researching and developing entrepreneurial skills. Business can only be improved if the entire socio-technical network of agriculture is involved in the process. Therefore, strategies are needed to promote and strengthen agricultural entrepreneurial culture and sustainability in rural areas. This chapter discusses the role of agricultural entrepreneurs in the development of agriculture, its scope, opportunities, importance and success stories of agricultural entrepreneurs.

Defining Entrepreneurship

Entrepreneurship refers to the capability to generate concepts and succeed with them. Innovation, which includes the capacity to embrace changes and take risks, along with the management of resources, constitutes essential elements in building a sustainable business. Entrepreneurship is responsible for employment, competitiveness and the ability to exploit any sector or business (European Communities, 2003). Entrepreneurship is a viable approach to upward mobility, as a 1% increase in entrepreneurship reduces the poverty rate by 2% (Singh, 2014). Entrepreneurs who run successful businesses are independent.

The word entrepreneurship means a dynamic process of progressive creation of assets (Shailesh al., 2013). These assets are produced by individuals who take the greatest risks of capital, time and career commitment to provide value for a product or service. It is possible that the product or service could either be new or unique or not, but the value must be instilled by the entrepreneur. Simply put, entrepreneurship is the use of energy to start and build a business (Mishra *et al.*, 2010). Thus, entrepreneurship is an exciting concept that is widely used and defined as a creative and innovative response to the environment (Chandramouli *et al.*, 2007).

It can also be defined as the process by which an agent manages to visualize and implement an idea, belief, service, product or activity (Bernier and Hafsi 2003). Stevenson *et al.* (1990) describe it as the pursuit of opportunities regardless of whether resources are sufficient or not. Other authors see it as an individual's efforts to make a vision a reality despite the probability of success. This may indicate that anyone can be an entrepreneur with the potential to develop a vision to create their own. In addition, it can be interpreted that entrepreneurship is a mutual phenomenon, where society as a whole can promote general well-being with the help of new products and services.

Agri-Entrepreneurship

Research shows that the agricultural sector is relatively more effective than any other sector in reducing poverty. Food prices fall due to increased crop productivity for the poor, leading to reduced poverty.

Agribusiness can generally be defined as sustainable, community-based, direct market agriculture. Sustainable agriculture refers to a systems-based approach to agriculture that emphasizes the interconnectedness of social, economic, and environmental dynamics. It is a useful combination of farming and trading and turns your farm into an agribusiness. This association of agriculture and business promotes agricultural entrepreneurs who innovate, identify markets and meet needs by developing different ways. The concept of agribusiness is similar to entrepreneurship in agriculture and describes the establishment of agribusinesses in agriculture and related sectors (Bairwa *et al.*, 2014a).

Entrepreneurship in agriculture can also be defined as the formation of a new type of economic organization to grow under conditions of risk and uncertainty in agriculture (Dollinger, 2003). Conversely, Gray (2002) defines an entrepreneur as an individual who runs a business to grow the business and the management skills necessary to achieve those goals.

Need and Importance of Agri-Entrepreneurship

Traditionally, farmers are not familiar with scientific agriculture and effective agricultural management systems. Therefore, they are unable to cope with delayed monsoon seasons, drought, crop debt, adulterated seeds and lack of fertilizers, resulting in them opting for suicide. Thus, agricultural management skills, technical and innovative entrepreneurial skills can nurture a well-educated agricultural entrepreneur who becomes a role model for all depressed farmers.

Agribusiness has the potential for social and economic development, such as job creation, poverty reduction, improved nutrition, health and overall food security in the national economy, especially in rural areas. Due to increasing unemployment and poverty in rural areas, entrepreneurship is urgently needed to increase productivity and profitability in agriculture. Agribusiness can serve as the definitive remedy to address this complexity, such as lightening the burden on agriculture, creating job opportunities for rural youth, managing rural-urban migration, increasing national income, supporting rural industrial development and cuts. reduce pressure on cities.

Agricultural entrepreneurship programs are critical for developing entrepreneurs and leaders in the agricultural sector worldwide (Bairwa *et al.* 2014b). Agricultural entrepreneurship is highly influenced by the economic situation, education, and culture (Singh, 2013). Agricultural entrepreneurship is important to the country's economy in the following ways (Sah, 2009).

- First, it helps smallholders achieve productivity and connect them to local, national and international markets.
- Second, it helps reduce food costs and provides a quality diet to the country's rural and urban poor.
- Third, it accelerates growth, diversifies income and develops business opportunities in rural and urban areas.

Entrepreneurial Skills and Characteristics for Agri-Entrepreneurship

Development of entrepreneurship is an effective human development that is forbidden to entrepreneurs. Developing an entrepreneur means instilling entrepreneurial skills, including the desired knowledge, higher technical, financial, marketing and management skills in the common person, as well as forming an entrepreneurial vision. Entrepreneurship development programs can be defined as a program that aims to help an individual develop entrepreneurship and create the skills and abilities needed to effectively manage their entrepreneurial role. In the literal sense, entrepreneurial competence is competence essentially necessary to perform tasks and activities in agriculture, which can be developed through study and experience.

The personal characteristics of the agricultural entrepreneur clearly influence agricultural entrepreneurship (Brockhaus and Horwitz, 1986; Nandram and Samson, 2000). In fact, entrepreneurs share common characteristics such as persistence, drive, drive, resourcefulness, problem solving skills, realism and goal orientation. An entrepreneur is one who recognizes an unmet need and takes the risk to fulfil it. In addition, other characteristics include productivity management, looking for new markets (Singh, 2013), self-criticism, leadership, market orientation and creativity.

Open farm entrepreneurs experience more problems in a dynamic environment (Hanf and Muller, 1997). According to Man et al (2002), entrepreneurial skills are classified into six key areas, which are e.g. opportunity recognition skills, relationship building, leadership, strategic skills, abstract thinking and problem-solving skills. Another author chose seven important aspects of success, which are organization and strategic planning, knowledge of the ecosystem, experienced and professional employees, value chain oversight, craft skills, ability to learn and seek opportunities, and creative personality traits (Lauweres, 2002). Management and organizational skills are a whole that a farmer could use to develop the business of the farm.

Small farmers have increased their competitiveness through producer organizations and technological innovations. Farmers had to adapt to new situations to face changing agricultural markets such as ICT, reduced agricultural subsidies, changing consumer demand and new food production standards (Lans *et al.*, 2013). Agribusiness is quite different from traditional business because agricultural modernization focuses narrowly on the highly specialized areas of farm management, efficiency and productivity.

Agricultural entrepreneurs must be innovative, curious, determined, long-term, visionary, hard-working, honest, risk-taking, proactive in identifying opportunities for success, integrity, and strong leadership and organizational skills (Bairwa *et al.*, 2014a). It can be said that the characteristics of an agricultural entrepreneur are a combination of characteristics and skills of an agricultural entrepreneur (Figure 6.1).

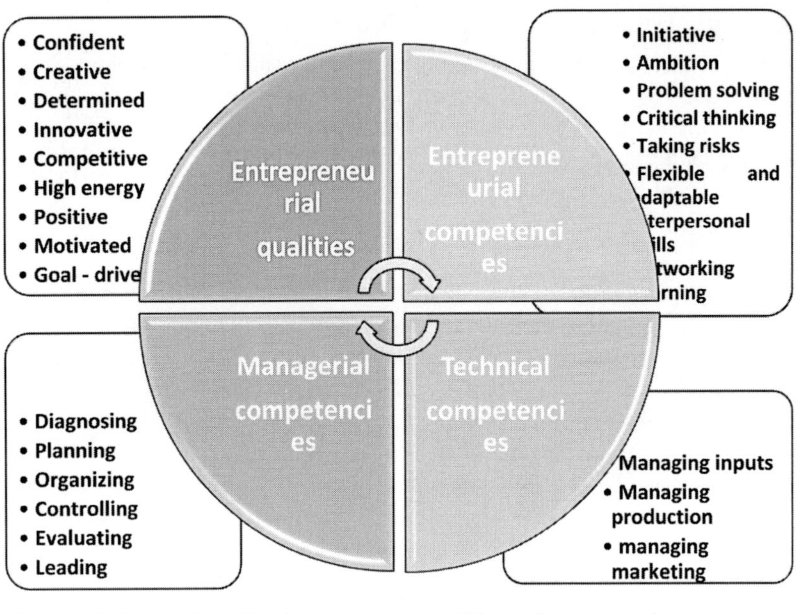

Figure 6.1: Integration of agri-entrepreneur qualities and competencies

McElwee and Smith (2012) recommended that agricultural entrepreneurs should be confident in their abilities to develop a successful business. Or we can say that they are intrinsically motivated to succeed. The characteristics most associated with agricultural entrepreneurs (Figure 6.2) include:

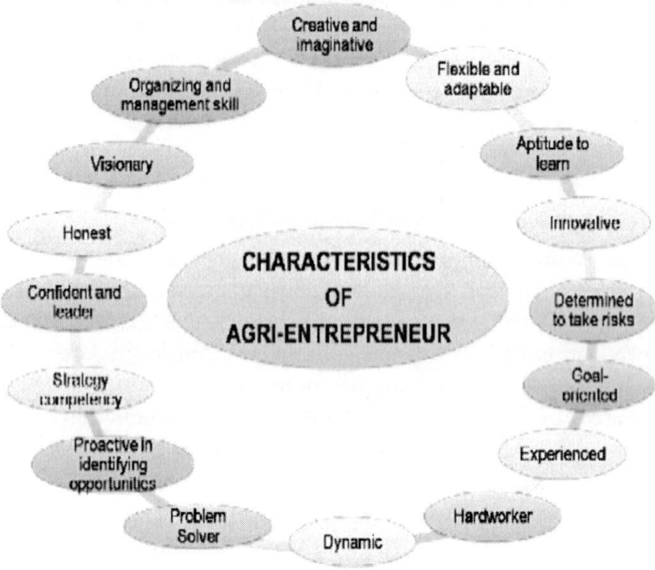

Figure 6.2: Characteristics related with agri-entrepreneurs

Entrepreneurship Development Opportunities in Agriculture

With the globalization of trade and agriculture and political reforms at the national level, the reach and opportunities of agriculture have greatly expanded, leading to extraordinary business interest in the sector. The world marvels and trusts in India's rapidly growing land market, which is crucial for building the country's business growth strategy. According to research, the rural market in India is bigger than the urban market.

Agribusiness has provided vast opportunities for value addition, packaging, retailing and export of agricultural products through advanced technology and management. A large part of India's population depends on agriculture and this region also supplies raw materials to various industries. Agricultural activity is likely to dominate the incremental growth of the Indian economy. In the current scenario, about a quarter of our fruits and vegetables spoil before reaching the consumer due to lack of proper infrastructure. However, due to the rapid changes in the nature of agriculture, the demand for competent and dynamic specialists has increased several times in recent years. Since the WTO system introduced policy reforms, opportunities and opportunities for the agricultural sector have increased. Great opportunities have opened up in industries such as packaging, supply of raw materials, processed agri-food production, export of agricultural products and other related industries. Increase in microfinance, relaxation of government regulations, availability of high technology, training and workshops in agriculture and allied sectors have changed the attitudes of highly skilled workers to choose self-employment in agriculture, increasing opportunities for agribusiness in India (Bairwa). *et al.*, 2014a). Pandey (2013) suggests many business areas in agriculture including dairy farming, farming, goat farming, rabbit farming, flower farming, fishing, shrimp farming, sheep farming, vegetable farming, livestock farming and farm forestry.

Possible opportunities for business in agriculture are:

- Agricultural Product Processing Units - No new products are produced here, only processing of agricultural products is done - for example, rice mills, dal mills, shell mills, etc.
- Agricultural Products Manufacturing Units - Here are brand new products made from agricultural products as the main raw material. For example, sugar factories, bakery, straw boards etc.
- Agricultural input production units - here production of goods either for mechanization of agriculture or for raising production buildings, for example fertilizer production units, food processing units, agricultural machinery etc.
- Agricultural Service Centers - These include repair shops and service centers related to the repair and maintenance of agricultural machinery used in agriculture.
- Various Areas - In addition, the following areas can be recommended for establishing agribusinesses such as setting up apiaries, fodder processing units, seed processing units, mushroom production units, commercial pest production units, goat nest clubs, organic fruits and vegetables. retail shops, bamboo plantations and jatropha cultivation.

Bansal (2011) suggested some areas of entrepreneurship in agriculture.

- Agriculture - Farming, Dairy/Poultry/Goat Farming, Fish, Rabbits, Vegetables, Flowers, Ornamentals, Palm Rose, Fodder, Agriculture, Agroforestry, Beekeeping, Mushrooms.
- Product marketing - Wholesale, retail, brokerage, transport, export, financing, warehousing, consulting.
- Marketing of Inputs - Fertilizers, Agricultural Chemicals, Seeds, Machinery, Feed, Poultry, Veterinary Medicine, Landscaping, Agricultural Credit, Ordering Service, Biomonitoring Units, Biotechnology Units.
- Processing - Milk, Fruits, Vegetables, Fish, Sugarcane, Cashew, Coconut, Poultry, Livestock, Tannery, Brewery.
- Brokerage - research and development, marketing data, quality assurance, insurance, energy.

Case- Vermicomposting Brings Better Yields and Returns - A Small Farmer Shows the Way

This is the story of a small farmer who opted to stand out from the usual resource-scarce farmers battling to survive in the degraded drylands. His profound 'love for life' has not only brought him a good living, but more significantly, it has motivated many other farmers to follow his example.

Chandranna, a young farmer, was previously known as 'Nursery Chandranna' and is now referred to as 'Vermicompost Chandranna. ' He has made Rs. 1. 4 lakhs from the sale of vermicompost and earthworms over the span of three years. It has now turned into a fairytale in the area where the average yearly income for a small farmer like him does not surpass Rs. 15000.

An ideal village, a less than desirable environment, Chandranna resides in Tumkurlahalli, a village comprising 650 households in Molakalmuru Taluk of Chitradurga District in Karnataka. The village is mainly made up of backward communities – 410 SC families, 100 Muslims, and 100 Lingayat families, with a total population of 3800. The village encompasses approximately 3322 ha, of which 15% is dryland and 3. 5% is irrigated using bore wells. The remaining 2695 ha, (81. 5%) consists of common land, which includes wasteland, communal grazing areas, and the 'reserve forest' where only occasional shrubs and bushes are present. The landscape, generally, features shallow red sandy soils. With boulders dispersed throughout, it is not an optimal village for profitable farming. With an average annual rainfall of less than 500 mm, farmers are compelled to take a risk with groundnut cultivation, the sole cash crop planted year after year. Over 30 years of planting groundnut exclusively has led to yield levels plummeting to a meager 8 q. /ha. While agriculture is hardly a profitable venture, most of the population continues to rely on agriculture and wage labor for their livelihoods. Naturally, the village experiences the migration of men for the majority of the year.

Small farmer with big hopes

In a village where agriculture seldom yields surprises, Chandranna's situation illustrates that strong interest and self-confidence can turn agriculture into a reliable venture. For, it is not a tale of immediate triumph but rather one of a methodical effort to take advantage of opportunities provided to the farmers in the village by various agencies.

Hailing from a low-income farming family, Chandranna has inherited 3 acres of dryland, of which one acre is unarable wasteland. Thus, wage labor was a more significant source of income for the family than agriculture on the two acres of land. His parents wished for their only son to pursue education. However, due to financial constraints, it was not feasible for him to go further than the pre-university stage. He was compelled to return and assist his parents in farming. Within the Karnataka Watershed Development (KAWAD) project, with AME Foundation serving as a resource agency, Chandranna became a member of a Self-Help Group (SHG).

The triggering point

In 2000, Chandranna attended a training on nursery cultivation at BAIF Institute of Rural Development, Karnataka (BIRD K), located in Tiptur. However, he was more eager to learn about vermicomposting, which was a concurrent training session being held for a different group of farmers at that time. He would participate in the group whenever he had the chance. He became enthusiastic about breeding earthworms and making vermicompost.

Upon his return from nursery training, his group received the chance to cultivate a nursery of 15,000 seedlings. The responsibility was assigned to Chandranna. From the year 2000 onward, Chandranna operated the nursery for three straight years. In 2003, his nursery was recognized as the finest in the watershed project, and he earned the nickname 'Nursery Chandranna'.

A humble start and an impressive rise His interest in vermicomposting persisted. With the limited knowledge he had acquired during the training, he attempted to breed local earthworm species in coconut shells. Unfortunately, they did not survive.

In 2003, Chandranna constructed four vermicompost pits measuring 6x3x3 cu. ft. with assistance from the KAWAD project. However, he was unaware of how to utilize the pits. A GUARD staff member later provided him with 2 kg. of earthworms, which cost him Rs. 300. Using the 2 kg. of earthworms, he produced 20 q. of vermicompost, which he applied to his 2 acres of ragi crop. Cultivating ragi was, in fact, an experiment in Tumkarlahalli as no one in the village had previously grown ragi. He harvested 14 q. from the 2 acres.

In 2004, he applied 6 q. of high-quality vermicompost and 2 tractor loads of FYM (2 tons) along with a bag of DAP over the same 2 acres. During this season, he grew groundnuts and achieved a yield of 20 bags totaling 9 q. of groundnuts.

Exploring tree-based farming systems and engaging with farmers who had successfully implemented composting and vermicomposting provided Chandranna with a wider perspective on sustainable agriculture. He gained additional knowledge about vermicomposting during his visit to another progressive farmer in the nearby village, B G Kere.

In the year 2005, Chandranna applied 6 q. of vermicompost to a one-acre PTD plot alongside a series of combined practices such as summer ploughing, treating seeds with bio agents (Rhizobium and Trichoderma), applying Gypsum (50 kg), utilizing a seed rate higher than normal (45 kg), and cultivating intercrops and border crops. The yield increased to 13 bags from one acre, resulting in 6. 5 q. of groundnuts. This was the highest yield recorded by a farmer from one acre over the last four years of AMEF's activities in the area. What stood out was the weight of each bag, which varied between 50 and 60 kg. While Chandranna's 25 bags weighed 13 q, his neighbor Tippeswamy's 40 bags only weighed 13 q. The trader purchasing the produce was astonished by this. In fact, the traders compelled Chandranna to empty the contents of the bags to verify that they did not contain stones. It was uncommon for a bag of groundnut pods to exceed 50 kg in weight. The consistent maturity of the pods and proper filling had significantly enhanced the quality of the groundnuts.

Chandranna did not cease his activities at producing vermicompost and utilizing it on his two-acre property. He commenced the sale of both earthworms and vermicompost starting in 2004. In that year, Chandranna sold 124 kg of earthworms at Rs. 150 per kg, which resulted in earnings of Rs. 18,600. Additionally, he earned Rs. 7,500 by selling 15 quintals of vermicompost at Rs. 500 per quintal. All together, he garnered approximately Rs. 26,100.

Motivated by an income greater than that obtained from groundnut, he amplified his production and sale of worms and compost in 2005. Throughout this endeavor, he faced some challenges and acquired valuable lessons. There was an instance when he packed 30 kg of earthworms in soil culture for selling, which unfortunately perished before the transaction concluded. Subsequently, he shifted to selling worms packed in cow dung. As the watershed project entered its final year and provided more vermicomposting pits to a considerable number of farmers, the demand for earthworms escalated. He earned Rs. 41,700 from the sale of 278 kg of worms (priced at Rs. 150/kg) and Rs. 11,500 from selling 23 quintals of compost at Rs. 500 per quintal. This led to a total earning of Rs. 53,200 for the year 2005. He further expanded the quantity of vermicomposting pits. Chandranna began searching for more crop residues and agricultural wastes. The four pongamia trees in his field, the biomass from the trees along the canal, and the dry eucalyptus leaves served as raw materials for his vermicompost pits. Acknowledging the necessity for cow dung in vermicomposting, Chandranna began maintaining a pair of bullocks, a cow, and 20 hens. The returns have been consistently increasing.

In 2006, despite experiencing an unprecedented drought-like situation not witnessed in the last 50 years, Chandranna still managed to earn Rs. 58,750 from selling 285 kg of worms and 32 quintals of vermicompost. His total earnings since 2003 reached Rs. 1,38,050. The actual profits could potentially be significantly higher. The Rs. 1. 4 lakhs he reported are documented in the accounts he maintains through receipts. His 'customers' primarily consist of self-help groups (SHGs) and farmers associated with numerous NGOs in the districts of Bellary, Chitradurga, Bagalkot, and Bijapur, who request bills. There are individual farmers purchasing compost or worms without asking for bills, making it impossible to document those transactions. Presently, he provides a special rate of Rs. 100/kg for SHGs, while others must pay Rs. 150.

Customers located nearby receive an additional after-sales service from Chandranna. He visits his customers' farms, and if the survival rate of worms is not up to standard, he offers them additional worms at no extra charge.

The beacon of hope for the hopeless

The well-known name 'Nursery Chandranna' has now been renamed to 'Vermicompost Chandranna'. A simple mud home is now being expanded with cement walls, accompanied by an increase in the number of vermicomposting pits in the backyard. He has already motivated many farmers in his village to explore alternative farming methods in general and adopt vermicomposting specifically. The self-driven farmers like Chandranna are exactly the type of catalysts that NGOs seek in order to transform small achievements into widespread movements. It represents just the right kind of encouragement that many resource-strapped farmers require to surpass their limitations and challenge the odds.

Source: AME Foundation (https://en.vikaspedia.in/viewcontent/agriculture/best-practices/agri-based-enterprises/case-studies-agri-enterprises)

Types of Enterprises

According to Uplaonkar and Biradar (2015), the various businesses in agribusiness are:

- Farm-level producers: Here, the family is perceived as a means to enhance output by employing significant levels of technology, assets and market demand.
- Service providers: There exist various types of service providers in the village setting. These encompass the lending and sharing of resources and the utilization of machinery like tractors, sprayers, planters, threshers, harvesters, and "dryers. " It also comprises scientific services such as constructing irrigation facilities, weeding, safeguarding plants, harvesting, threshing, transporting, storing, etc. , along with associated services in husbandry, breeding, vaccination, disease diagnosis and treatment, as well as animal feed, minerals, grains, etc.
- Producers of inputs: There are numerous thriving businesses that need considerable resources, and these resources can be generated by local entrepreneurs at the community level, such as biopesticides, soil conditioners, biofertilizers, vermicompost, assorted vegetable crops, fruits, ornamental plants, root crops, growing media for potted plants, livestock feed for concentrated feed production, farming tools, irrigation systems, mineral supplements, and complete animal feeds. In addition, there are opportunities in fisheries, agriculture and poultry farming.
- Processing and Marketing of Agricultural Products: Well-structured management of post-production activities necessitates well-trained and experienced personnel and greater investment. Such organizations may be overseen by non-governmental entities like cooperatives, public service firms, or communities. The most effective instances are milk cooperatives, sugar

cooperatives, and fruit cooperatives.

Micro, Small and Medium-Size Enterprises

Small-scale farmers can address issues like underemployment and the necessity for new job opportunities through entrepreneurship, as they are positioned to create value-added agricultural goods intended for sale in local markets. Shoppers seek transparency within the food supply chain, which promotes the localizing of agro-food systems (LAS) (Requier-Desjardins *et al.* , 2003). Farmers in rural areas can rival larger companies by establishing a local production system (LPS) that manages significant volumes of local agricultural produce (Requier-Desjardins *et al.* , 2003). These clusters enhance the productivity of production operations for micro/small/medium enterprises (MSMEs) and lower business expenses.

Microenterprises are defined as businesses with one to nine individuals; small businesses employ between 10 and 49 staff members, while medium-sized enterprises have a workforce ranging from 50 to 250 employees (Andersén, 2012). Furthermore, due to the division of labour, SMEs establish both cooperative and competitive interactions within these clusters (Trienekens, 2011). The division of labour enables SMEs to be agile and modify their products or services in response to market requirements. Clusters also enhance information exchange, dissemination, and innovation throughout firms.

Value-Added Agribusiness

New possibilities for agricultural entrepreneurs have arisen as a result of consumer preferences for locally produced and value-added agricultural items (DeLind, 2002). Value-added items encompass organic vegetables, BST-free cheeses, unbleached flour, gluten-free foods, and heirloom apples. The importance of value-added products lies in the fact that their sale generates additional revenue for farmers and bolsters local and regional economies (Vogel, 2012). Value-added agricultural ventures include technical services, rental of farming equipment, agritourism, forest products, and community-supported agriculture. Therefore, agricultural enterprises typically generate more income compared to non-agricultural businesses. There has been a rise in consumer demand for locally grown and organic crops (Ikerd, 2011). Agricultural businesses need to accurately label their products to ensure consumers are willing to pay the best price for them. Overall, such sustainable agricultural enterprises hold significance regarding ecological integrity and economic advantages (Ikerd, 2008). Farms that utilize sustainable agricultural practices have demonstrated better soil and moisture retention, reduced erosion, and fewer economic losses in comparison to those who do not engage in sustainable agriculture (Altieri *et al.* , 2012). Farmers in developing nations also have a substantial opportunity to profit from consumers in developed nations who are prepared to pay higher prices for products they associate with sustainability (Elliott, 2013). Nonetheless, there are challenges to developing value-added products, including insufficient market knowledge and resources, along with infrastructure and institutional obstacles. It is crucial for an agricultural entrepreneur to assess how to enhance the value of their product and manage the associated costs. According to Lingelbach *et al.* (2005), agricultural entrepreneurs in developing countries face limited access to credit and have networks that ensure product marketing. Ultimately, agricultural entrepreneurs seek to identify opportunities to advance their value-added endeavors, such as introducing new products or services to consumers, enhancing the distribution network's position, or forming partnerships to promote their business. A direct relationship evolves between the consumer, the farmer, the farm, and the bioregion, fostering a sense of place that connects the consumer, the producer, and the location, which is essential to the social self (Feagan, 2007). This strategy has proven effective at both local and national levels.

Scope For Agri-Entrepreneurship in India

Bansal (2011) offered a comprehensive summary of the magnitude of Indian agribusiness and this can be easily understood from Figure 6.3.

- India is endowed with a diverse range of aggressive climates that support the cultivation of temperate, subtropical, and tropical agricultural goods.
- The need for agricultural supplies such as feed and feed, synthetic fertilizers, biofertilizers has increased.
- Biotechnology applications in agriculture are a boon for seed production, industrial use of biopesticides and microbes for various products.
- Economic growth accelerates through the implementation of exports. According to the World Trade Organization (WTO), India possesses significant potential to restore its existing status in global trade concerning both raw and processed agricultural goods. At present, processing occurs solely at the primary stage, and the improvement in living standards enhances the opportunities for secondary and tertiary processing of agricultural products.
- The wide coastline and inland water bodies should be used for the cultivation of marine and inland fish. The cultivation of ornamental fish is already becoming popular because of its rising aesthetic appeal among Indian citizens.
- The wealth of livestock provides ample opportunities for the generation of meat, milk and dairy products, poultry, etc.
- Forest resources can be used to produce forest by-products.
- Beekeeping and beekeeping can be widely practiced in India.
- By enhancing the production technology, the domestic use and export of mushroom production can be elevated.
- Farmers ought to be motivated and educated in organic farming since organic farming holds the highest potential in India as it utilizes fewer pesticides and inorganic fertilizers compared to the industrialized nations worldwide.
- Significant prospects exist for the production and marketing of biopesticides for the protection of crops.
- Because of the productivity plateau of high-yielding varieties, seeds, hybrids, and genetically modified plants possess the most promising future potential in India.

- Micro-irrigation systems and labor-saving agricultural tools for farming undertakings such as sowing, planting, and harvesting hold promise in the future owing to decreasing groundwater levels and workforce shortages.
- The cultivation of vegetables and flowers in greenhouse environments can be used for export.
- Human resources educated in agriculture and associated fields will be obtained for the aim of the agricultural extension system in relation to the diminishing financial resources of the state and the decrease of the current agricultural extension staff of the state as consulting services.
- Ultimately, as a result of a rise in agricultural production, there are employment possibilities in marketing, transportation, cold storage and warehouses, credit, insurance, and logistics.

Figure 6.3: Scope of Agri-entrepreneurship

Barriers to Agri-Entrepreneurship Development

According to Uplonkar and Biradar (2015), the advancement of agricultural business is essential for enhancing production and productivity. Nevertheless, the success rate in India is quite low because of the following reasons:

- Agriculture primarily serves as the means of survival for the majority of farmers. For an uneducated farmer, transforming his farming into a business is challenging because of insufficient knowledge, funding, technology, and market connections.
- It is essential to educate farmers about the advantages of these services since they are not informed about them.

- Free service provided by government organizations to promote services should be stopped. Because many farmers, especially political leaders, feel that the government has a responsibility to offer extension and technical advisory services to farmers. Furthermore, the services provided by these organizations do not extend to small farmers, particularly those residing in isolated regions. This notion of complimentary service causes farmers to hesitate to utilize the paid services provided by local self-employed.
- A regular reservation service for FIEs, communication with marketing agencies, suppliers and research facilities connected to the advancement of contemporary technology is necessary.
- They encounter various legal limitations and challenges, and private business owners involved in such enterprises often disregard these regulations and disrupt the fairness of trade.
- Finally, organizations experience their own risk when they make large investments and adopt modern technologies that affect profitability. Thus, farmers lose interest in their companies and managers.

Other authors have identified and reported the following barriers to agribusiness. First, agricultural entrepreneurs must deal with changes in consumption patterns, desired products, distribution systems, emerging technologies, and the structure of the industry (Boehlje *et al.*, 2011). Nonetheless, there exists a distinction among the locations of production, processing, and consumption of agricultural goods. Furthermore, there is an increased level of risk and uncertainty in decision-making. There is a necessity for innovation, development, and the introduction of new technologies that facilitate enhanced production efficiency and profitability. Additional challenges consist of transformation, competition, changing industrial frameworks, and climate change. Factors such as geographical location, farm size, topography, surging prices, product supply, low and fluctuating incomes, environmental costs, poverty, uncertain livelihoods, lack of technology, socio-political instability, natural disasters, environmental degradation, gender discrimination, and high startup costs for businesses, along with difficulties in obtaining credit during the startup phase, contribute to these challenges (McElwee and Smith, 2012; Pettinger, 2010; Richards and Bulkley, 2007). Multinational corporations seek countries with more lenient foreign direct investment regulations as this enables them to enhance market access and exert control over market activities (Land Matrix, 2015). Consequently, this diminishes opportunities for local agricultural enterprises that lack access to public media and financial resources. Gender discrimination is evident in developing nations (Stupnytska *et al.*, 2014). Shifting economies often favor the export of agricultural goods, thereby exacerbating poverty by displacing farmers and limiting food availability for residents of those countries, particularly in rural areas (UNIDO, 2015; McMichael, 2005). This impacts the capability of local consumers to procure food and sustain their families.

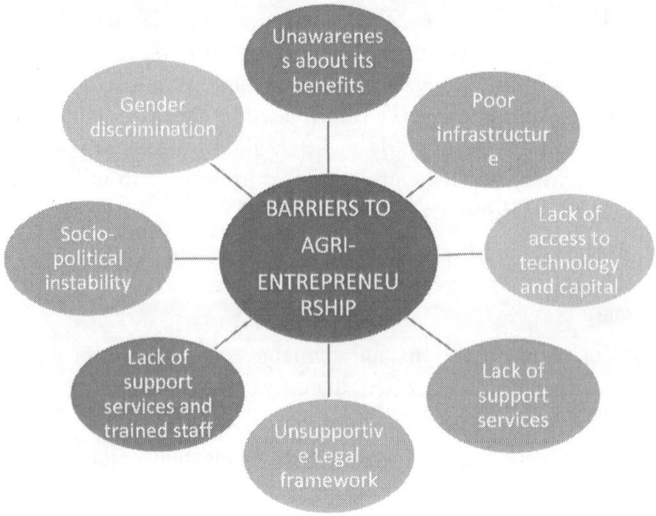

Figure 6.4: Barriers to the Agri-entrepreneurship development

Motivation for Agribusiness Start-Ups

As mentioned earlier regarding the challenges and issues encountered by agricultural entrepreneurs, the question emerges, what motivates them to establish their own enterprise? Prior to launching this venture, these entrepreneurs assess alternative job prospects and their influence on their family (Lan Matrix, 2015). The primary driving force for agricultural entrepreneurs is to enhance household income and broaden the core agricultural activity to involve more family members (Agnete *et al.*, 2003). In other terms, agribusiness can serve as a means of achieving a comfortable livelihood. Additionally, it benefits individuals who lack other professions. Some seek to leverage unique resources and employ them to gain a competitive edge in the market. Smallholders can remain economically sustainable by scaling up their operations and marketing value-added products (Lowson, 2002; Haugen and Vik, 2008). The uniqueness of the product must be preserved to boost the chances that consumers will purchase their higher-priced items and generate sufficient profit.

Training and Education Programme for Agribusiness (Uplonkar and Biradar 2015)

India educates over 12,000 graduates annually in agriculture and related areas, yet only a small number secure jobs. Consequently, policy makers recognize an urgent requirement to tackle this critical dilemma through effective educational planning. The Agricultural Clinics and Farm Business Centers Central Sector Program (AC and ABC) was initiated on 9 April 2002 with the objective of enhancing and broadening public initiatives by assisting qualified agricultural professionals in establishing farms and offering value-added counsel. providing services to farmers at their homes and also creating self-employment opportunities for farmers. Under the guidance of the Ministry of Agriculture, Government of India and the National Institute of

Agricultural Extension Management (MANAGE), Hyderabad is executing a scheme of agricultural clinics and agribusiness centers. The aim of the system is to enhance the current extension network to expedite the agricultural transfer process and reinforce the supply of production inputs and services. Participants trained in such initiatives consist of agricultural graduates, postgraduates, and individuals holding degrees in agriculture and related disciplines who can establish their own agricultural clinics and agribusinesses and offer professional extension services to farmers. Therefore, the system provides self-employment prospects for agricultural graduates and improved farming techniques for farmers. Specialized training is provided at no cost to eligible agricultural graduates. This training encompasses various facets of entrepreneurship and business management. Centre for Entrepreneurship Development (CED) Hyderabad is among the accredited Nodal Training Institutes that provides a two-month training program.

Other agencies and programs

- National Bank for Agriculture and Rural Development (NABARD) - NABARD was established by RBI in July 1982 to give full attention to agriculture, small scale household and agro industries in rural areas. NABARD is successfully managing all the activities of RBI related to rural development and agriculture related activities.
- Agricultural Technology Management Agencies (ATMAs) - ATMAs are operating at the district level to utilize the expertise of established Agri-entrepreneurs in delivering value-added extension services to farmers through a Public Private Partnership approach.
- Panchayatmandi (Agri-Mandi) - The concept of Panchayat mandi aims to lessen the influence of middlemen and merchants. Achieving this is feasible only if the Zilla Panchayat collaborates with the National Marketing Boards and APMC. (Agricultural Produce Marketing Committee).
- State Agricultural Marketing Banks (SAMBs) - Such banks are established to actively monitor food and oilseed markets in major urban markets.
- NCOSAMB (National Council for State Marketing Board) - The Government of India gives grants to states to set up modern training institutions and NCOSAMB manages such training programmes.
- State Trading Corporation (STC) - STD is the main international trading center of the Government of India. The company deals with large international transactions.
- Farmers Market - This is an original initiative of some state governments, commonly known as Uzhavar santhai (Tamil Nadu), Rythu Bazaars (Andhra Pradesh), Apna Mandi (Punjab). This type of market explores the huge potential of agricultural entrepreneurship education to develop entrepreneurial skills and is facilitated by KVK.
- Domestic and Export Marketing Intelligence Cell (DEMIC) - This indirectly

supports the entrepreneurial and decision-making skills of agricultural entrepreneurs. In addition, it also works as a commodity price forecaster.

- Agricultural Biotechnology Agency for Rural Employment and Development (ABARD) - This is a scheme of KAU (Kerala Agricultural University) and provides technological training.
- Pasumai Padai - Established by the Pondicherry Government, it focuses on the interests of farmers and engages in activities such as rental of agricultural machinery and equipment.
- Wine Growers Association, Maharashtra - It consists of 150 cooperatives and 500 growers. This association functioned as an international supplier of high-quality grapes.
- ITC-e-choupal - It provides weather and price information to farmers, creates a direct link between business goals and social goals and imparts knowledge of business and entrepreneurship skills.
- The Agribusiness Incubators-ABI initiative was initiated in 2003. It represents a collaboration between ICRISAT and India's Department of Science and Technology (DST). It endorses agricultural technologies created by ICRISAT, additional R and D centres of excellence, universities, and various institutions.

Examples Of Successful Agri-Entrepreneurs (Cases)

Success stories of agribusiness entrepreneurs have been documented from different sectors and states, highlighting their increasing significance in agricultural extension (Srinivas IAS *et al.* , 2014). This chapter showcases a selection of curated success stories. This will raise awareness among National Agricultural Universities, Banks, NABARD, Krishi Vigyan Kendras (KVK), and ATMAs regarding potential enhancements of the AC (Agri-Clinics) and ABC system for agricultural extension. This would motivate and support more upcoming agriculture professionals to build a successful career for themselves.

1. **Apple Planet Agri- Clinics:** Mr. Bukhari completed AC (Agri-Clinics) and ABC (Agri-Business Centres) training at the Indian Society of Agribusiness Professionals (ISAP) in Srinagar in December 2012. He made the decision to establish an agricultural clinic to assist the farming community and promote self-employment. In Jammu and Kashmir, the primary crop is fruit, particularly apples, which necessitate multiple plant protection sprays along with various fertilizers. He offers a follow-up service and advises on the appropriate amounts of pesticides, fertilizers, seeds, and other inputs to enable farmers to cultivate their crops profitably and sustainably. Mr. Bukhari assembled 500 farmers and brought them together in one association called "Agro-Clinics of Apple Planet".
2. **Xcell Breeding and Livestock Services pvt. ltd:** Dr. Gajendrakumar Kantilal Bamania, a 33-year-old veterinarian with a Master's degree in international agribusiness, has enrolled in the AC and ABC training course at the International

School of Public Administration (ISPL) in Ahmedabad. He chose to expand his business in rural regions, where the demand for veterinary services surpasses that of urban areas. He established the firm "Xcell Breeding and Livestock Services pvt. Oy. " He also created software to oversee the processing services provided to dairies and set up a comprehensive laboratory. It provides various services to farmers including semen sales, livestock procurement, livestock sales, livestock registration, artificial breeding products, livestock counseling, livestock semen sales, and distribution of AI accessories such as tags, enclosures, gloves, and other agricultural products.

3. **Mobile Agriculture School and Services (MASO)**: Shri Vijay Bharath initiated the Mobile Agriculture School and Services (MASS) to assist farmers in isolated villages of Jharkhand and Bihar in utilizing modern agricultural technology directly at their homes. He received training through the AC and ABC system at the "Society for Rural Industrialization" facility. MASS functions as a mobile bus service. This bus is outfitted with audiovisual tools that provide details on current farming practices and market data. The bus delivers seeds, fertilizers, and plant protection products to farmers at a lower cost.

4. **Ocean Foods:** Shri. Uday Veer Singh Rana from Haryana participated in the two-month AC and ABC training course at the Indian Society of Agricultural Professionals (ISAP), Haryana. Throughout his training, he learned the scientific approach to honey processing. He subsequently established a honey processing business named "Ocean Foods. " He started to experiment with honey processing in an artistic manner. The processing facility is equipped with excellent storage capabilities and proficient employees for bottling, packing, labeling, etc. The firm earns15 lakhs in net profit in the first year and total sales of Rs. 80 lakhs.

7

Entrepreneurial Development Models

The term entrepreneur is derived from the French word entreprendre, which translates to "to undertake. "In the context of business, this refers to the initiation of a business. According to the Merriam-Webster dictionary, an entrepreneur is defined as an individual who organizes, directs, and takes risks in a business or enterprise. An entrepreneur can be described as "a person who establishes, organizes, and directs any enterprise, particularly a business, typically involving significant initiative and risk. " Peter F. Drucker (1985) characterizes an entrepreneur as a person who continuously seeks change, responds to it, and leverages it as an opportunity. Innovation serves as a unique tool for entrepreneurs, enabling them to perceive change as an opportunity for launching another business or service. An entrepreneur is often regarded as an innovator – a creator of new ideas and business processes. Essential leadership skills and effective team-building abilities are frequently considered crucial traits for successful entrepreneurs. Economist Robert Reich (2002) regards leadership, management skills, and team building as vital attributes of an entrepreneur. The realm of entrepreneurship is influenced by a variety of combinations of socio-economic, psychological, cultural, and additional factors, including caste/religion, family background, educational level, level of insight, professional experience, migration capabilities, entrepreneurial spirit, nature of the business, investment capacity, and ambition/moderation. Entrepreneurship is often defined as "the ability and desire to develop, organize, and manage business activities and associated risks for profit".

It is an effort to generate value by identifying business prospects. Essentially, it involves communication and management actions to mobilize financial and physical resources. A business may function within a corporate ecosystem, which frequently encompasses:

- Government initiatives and services that encourage entrepreneurship and assist entrepreneurs and startups.
- Organizations that are not part of the government, such as small business associations and groups that offer advice and support to entrepreneurs (e. g., through business centers or online platforms).
- Advocacy groups for small businesses that campaign towards government to enhance support for entrepreneurship programs and to promote laws and regulations that benefit small businesses.
- Commercial tools and facilities (e. g., business incubators and seed funding boosters).

- Programs for entrepreneurship education and training provided by schools, colleges, and universities.
- Funding options (e. g. , bank loans, venture capital funds, and grants from public and private sources).

Relationship Between Small Business And Entrepreneurship

The word "entrepreneur" is frequently mistaken for or used as a synonym for "small business. " While the majority of entrepreneurial ventures begin as small businesses, not every small business can be classified as an entrepreneur in the genuine meaning of the term. Numerous small businesses operate as sole proprietorships with just one owner or a handful of employees, and many of these small businesses provide a pre-existing product, process, or service and do not aim to expand.

In comparison, entrepreneurship provides a creative product, procedure, or service, and the entrepreneur generally aims to grow the business by hiring more staff, pursuing overseas sales, and so forth, a process supported by venture capital and angel funding. An effective entrepreneur possesses the capability to steer the business in a beneficial direction through adequate planning, adjust to shifting circumstances, and recognize their strengths and weaknesses.

Entrepreneurial Characteristics

An entrepreneur is someone who takes the chance of launching their own business for an anticipated benefit. He is the individual who originates an idea, creates a strategy, arranges resources, and executes the strategy to reach his objectives. Becoming an entrepreneur necessitates specific traits and abilities that are frequently developed through education, diligent effort, and careful planning.

Desire to succeed: An entrepreneur possesses a deep aspiration to achieve success in life. Their ambitions extend beyond merely accomplishing a single objective; instead, they are always striving for greater aspirations.

Vision: An entrepreneur holds a vision and imagines the methods and pathways to realize that vision. In this process, he envisions market needs, the social and economic landscape, and the technological context, from which he anticipates the future of his business.

Technical Knowledge*:* An entrepreneur possesses comprehensive understanding of every technical facet of his enterprise - whether it's technological, operational, financial, or market dynamics. Entrepreneurs seek to be informed about how things are progressing. They require time and initiative to explore the unfamiliar.

Novelty*:* Entrepreneurs do not adhere to conventional guidelines. They possess an ongoing urge to introduce innovation into the current marketplace. They envision answers to challenges that propel the development of new products and concepts.

Independence*:* An entrepreneur's longing for authority and assessment frequently complicates working within a structured setting. An entrepreneur requires autonomy in their tasks and choices. They do not adhere to existing regulations but establish their own guidelines and future.

Reasonable risk: Risk is a critical and fundamental component of business. Begley and Boyd determined that entrepreneurs who engage in moderate risks achieve a greater return on their capital compared to those who either avoid risks or undertake excessive risks. The entrepreneur embraces the unpredictability of the future and mitigates risks through research, planning, and skill enhancement. The entrepreneur ensures rent for the landlord, salary for the employees, and interest for the investor, with the expectation of earning more from the outlays in the future.

Leadership: Gary Yukl (1994) describes leadership as "the process of influencing others to comprehend and concur on what must be accomplished and how, along with the process of enabling individual and collective efforts to attain shared objectives. "Entrepreneurs exhibit leadership traits. They possess strong planning abilities, organizing skills, motivational capacities, effective communication skills, sound decision-making skills, a proactive approach to executing plans, and a focus on outcomes.

Communication skills: An entrepreneur needs to cultivate a positive relationship with his customers to achieve profits and secure their trust in his products/services. He must also uphold a positive relationship with his employees to operate the business successfully.

Diligent: Entrepreneurs exhibit a considerable amount of enthusiasm to launch a business and achieve success. They are occasionally referred to as workaholics. For them, work is akin to worship. They perpetually seek success and understand that there is no replacement for diligent effort.

Perseverance: True entrepreneurs encounter obstacles, err in their decisions, face critiques, and manage financial, familial, or stressful challenges. Yet, they continue to cling to their aspirations for their project to succeed. They maintain a strong sense of optimism regarding their vision. They perceive issues as chances for growth and challenges.

Confidence: Entrepreneurs have faith in themselves. Their self-assurance eliminates any uncertainties they may have.

Flexibility: Entrepreneurs need to be adaptable to adjust to shifting trends, markets, technologies, regulations, and the economic landscape.

Types of Entrepreneurs

Entrepreneurs can be categorized across various dimensions, a few of which are outlined below:

A. Classification on the basis of Ownership

1. **Founder or "Pure Entrepreneurs"**: These are people who establish their own company. They develop a business strategy and subsequently strive to ensure the strategy succeeds. For instance, Dhirubhai Ambani of Reliance Group.
2. **Second generation family business operators:** These are people who received the business from their fathers and ancestors. For instance, Mukesh Ambani and Anil Ambani, the two offspring of Dhirubhai Ambani of Reliance Group,

are currently separated into two entities - Reliance Industries Limited and Reliance-Anil Dhirubhai Ambani Group.

3. **Franchises:** This is a business framework in which the parent owner grants permission to use their brand and "established methods" to the franchisee for a continuous payment. For instance, NIIT transferred its franchise operations to regional operators following a detailed evaluation and adequate training.
4. **Owner-Manager**: When an individual purchases a business from a creator and subsequently dedicates their time and resources to it, they are referred to as owner-managers. For instance, Sabeer Bhatia and Jack Smith established Hotmail, which is now owned by Microsoft and is called Outlook. com.

B. Classification based on Personality Traits

1. **Innovator-Designer:** This kind of entrepreneur possesses a strong drive to succeed. The sole motivating factor for them is the aspiration to accomplish something meaningful in life, the aspiration to leave a legacy in society, the aspiration to demonstrate their capabilities. They do not require outside encouragement, but instead are guided by their own initiatives. These entrepreneurs see the potential for developing innovative ideas, new technologies, new markets, and new organizations.
2. **Induced Entrepreneur:** This kind of entrepreneur is drawn or inspired to establish a business by various external factors. External factors may include, for instance, government regulations, joblessness, familial assistance, supportive organizations, etc.

C. Classification based on the type of Business

1. **Industrial entrepreneur:** An industrial entrepreneur is an entrepreneur who creates a product. He discovers the requirements and desires of the customers and produces items based on their requirements and desires.
2. **Commercial entrepreneur:** A commercial entrepreneur is a business person who is not focused on producing goods. He recognizes possible markets, generates demand, and engages buyers in purchasing the product.
3. **Entrepreneur:** An entrepreneur is an individual who shows innovation in structuring and overseeing a business venture that is registered under any legislation that grants it the designation of a distinct legal entity.
4. **Agricultural entrepreneur:** Agricultural entrepreneurs are individuals involved in enterprises associated with agricultural activities, including seeds, fertilizers, pesticides, agricultural machinery, and other agricultural supplies.
5. **Social Entrepreneur:** A social entrepreneur is an individual who initiates a project to address various social issues like illiteracy, child marriage, drug addiction, environmental challenges, terrorism, and more. Frequently, these entrepreneurs do not aim to make a profit.

Role of Entrepreneurs

The role of an entrepreneur in contemporary production resembles that of a theater director. Current economic growth is intricately linked to production, which is inherently complicated. The involvement of an entrepreneur in today's economic progress encompasses at least three dimensions:

- The entrepreneur organizes various factors of production. This entails collecting factors and guaranteeing the optimal combination of them for the production process.
- The entrepreneur assumes risks. This is a crucial responsibility for an entrepreneur, since the level of profit he earns is directly related to his own risks. Risks are typically founded on demand prediction.
- Lastly, the entrepreneur brings about innovation. Innovation is distinct from invention. Invention refers to the work of scientists. Innovation involves the commercial use of an invention. As an innovator, the entrepreneur takes on the role of trailblazer and leader of the organization.

Importance of Entrepreneurship in Economic Development

Every nation aims to attain optimal economic growth, which significantly relies on its human resources. Nevertheless, human resources on their own do not create economic growth; there needs to be vibrant entrepreneurs. Entrepreneurship is among the most vital inputs for the economic growth of the nation. Entrepreneurs are essential not just for the progress of the nation's industry but also for agriculture and the service industry. The key functions of entrepreneurs in the nation's economic growth are outlined as follows:

1. **Employment creation:** increasing unemployment, particularly among educated individuals, is a pressing issue facing the nation. When one hundred individuals start their own businesses, they not only generate one hundred jobs for themselves but also hire numerous others. These businesses expand over time, thereby offering both direct and indirect employment opportunities to many more people. Consequently, entrepreneurship represents the most effective means of combating the problem of unemployment.
2. **Promotes capital formation:** Entrepreneurs facilitate capital formation by utilizing the unspent savings of the public. Both personal and borrowed capital are utilized to set up businesses. This type of economic activity results in the generation of additional value and wealth, which is crucial for the industrial and economic progress of the nation.
3. **Decay of economic power:** When a community generates a limited quantity of entrepreneurs, the industry expands into several major corporations because of insufficient competition. This results in the accumulation of wealth within a small number of families. This situation can have significant social and national repercussions. As the count of entrepreneurs rises, a substantial portion of national wealth is shared among a greater number of entrepreneurs, which

spreads out wealth.

Small business endeavors also allow underrepresented groups such as women, SC, ST, and OBC to achieve their financial aspirations via agro-based rural industries and handicraft-focused cottage industries. Because there are no obstacles to obtaining education, entrepreneurship becomes an even more appealing career choice for these marginalized groups. This allocation of wealth fosters genuine socialism and contributes to a healthy economy.

4. **Balanced regional development:** The swift progress of trade guarantees equitable regional development. As new entrepreneurs expand quickly, rising competition in urban areas is driving them to establish their enterprises in smaller towns distant from major cities. The expansion of industry and commerce in these locations results in a significant amount of public goods like roads, transportation, healthcare, education, entertainment, and more. It fosters the progress of underdeveloped regions and consequently aids in achieving balanced regional development.
5. **Developing backward and forward links:** Every new enterprise generates chances for input providers (referred to as backward links) and output sellers (known as forward links). These immediate links establish new relationships. Consequently, entrepreneurship enhances the degree of economic activity via a chain reaction.
6. **Financial Independence:** Entrepreneurship is crucial for a nation's self-reliance. Businesspeople ship goods and services on a large scale and generate valuable foreign exchange for the nation. This approach to substituting imports and promoting exports aids in securing the country's economic autonomy.
7. **Reducing youth unrest:** Many issues concerning youth and social tensions arise from the absence of productive employment for young people. In a shifting landscape where we encounter the challenge of reduced chances for paid employment, entrepreneurship serves as an alternative to wage work. The government needs to direct young individuals with entrepreneurial qualities from salaried jobs to entrepreneurial pathways. This alternative path via entrepreneurship might assist the nation in reducing social tensions and unrest amongst the youth.
8. **Increase in national income:** National income is made up of products and services created and brought into the country. Products and services that are created are meant for consumption within the country as well as for export purposes. As the population and quality of life improve, domestic demand increases. Additionally, export demand is rising for various reasons to satisfy the needs of higher imports. An increasing number of entrepreneurs must react to the escalating demand for goods and services. Hence, entrepreneurship contributes to the growth of national income.
9. **Increase in the standard of living:** The rise in people's standard of living is intrinsic to the nation's economic progress. Business owners are crucial

in enhancing the living standards of individuals. They establish industries that resolve shortages of vital goods and introduce new products into the marketplace. They also implement the newest innovations in the manufacturing of various goods and services at reduced costs. This enables individuals to access higher quality and a broader selection of goods at lower prices, resulting in an elevation of their standard of living.

Although entrepreneurship contributes to economic growth, the reverse is also accurate, meaning that economic development fosters the advancement of entrepreneurship as well. A growing economy creates a favourable environment for business to thrive.

Predictors of Entrepreneurial Success

The achievement of a company is shaped by various elements, including the organizational, marketing, and interpersonal approaches taken by the entrepreneur. Below are some key factors that forecast entrepreneurial success:

1. **Methods**
    - Develop corporate strategies, encompassing growth and survival strategiesRetention of human resources (recruitment and retention of talented employees and managers)
    - Guarantee the accessibility of essential materials (e. g., raw materials utilized in production, etc.)
    - Taking advantage of a company's distinct competitive strengths
    - Ensure effective organizational planning, proficient management, and organizational alignment
    - Alignment with the culture of the community
2. **Market**
    - Employing business-to-business (B2B) or business-to-consumer (B2C) frameworks
    - Strongly growing market
    - Focus on customers or markets that have not yet been utilized or overlooked by others.
3. **Team**
    - A significant team that is diverse in gender and race with varied skills
    - Experience in management before the implementation phase
    - Professional experience in the start-up industry
    - Full time job before new company
    - Previous business experience
    - Complete involvement in the new organization
    - Driven by various objectives, not solely financial gain
    - Connections within the community and breadth of business relationships

4. **Company**
   - Composed business plan
   - Centralize on a single cohesive product line or service line
   - Competition relying on a criterion apart from price (e. g. , quality or service)
   - Timely, regular, vigorous, and well-directed marketing
   - Rigorous financial oversight
   - Ample initial and expansion funding
   - Business framework, not a personal enterprise

Risks of Being an Entrepreneur

Entrepreneurship revolves around embracing risks, as nothing is guaranteed in the business world, and an entrepreneur must take risks to earn a profit. These risks can occasionally produce successful outcomes and, at times, result in losses. Listed below are some of the major risks encountered by an entrepreneur.

1. **Operational risk:** Since the entrepreneur holds complete responsibility for the business, the success or failure rests solely on him. An entrepreneur needs to be mindful of production expenses and must guarantee that neither time nor money is squandered to establish and sustain efficient production.
2. **Strategic risk:** New participants or rivals frequently emerge as effective alternatives to the offerings of an unfamiliar business. As a result, a business owner needs to implement a superior, imaginative, and inventive approach to attain lasting success in the marketplace.
3. **Technological risk:** In a new endeavor, there are numerous unforeseen changes that need to be managed efficiently to sustain the business. Innovative and more effective technology frequently renders the old ones obsolete. As a result, an entrepreneur must keep up with technological advancements and adjust to evolving technologies.
4. **Financial risk:** Numerous new companies tend to earn little income initially, which means the entrepreneur has to assume income risk during the early phases. Occasionally, government regulations, tax legislation, and changing exchange rates can significantly impact business earnings.
5. **Loss of free time:** It is common for entrepreneurs to put in significant overtime to ensure their business thrives. This is particularly significant during the early stages of start-up. Consequently, there is a shortage of free time for rest and leisure.

Factors Affecting Entrepreneurship Development

Business is impacted by four distinct factors: economic development, culture, technological advancement, and education. These circumstances can produce both beneficial and adverse effects on the growth of entrepreneurship. Positive influences promote and foster the emergence of entrepreneurship, while negative influences generate an atmosphere that obstructs it.

Economic Factors

Business is influenced both directly and indirectly by the economic environment. This is likely due to individuals becoming entrepreneurs out of necessity when alternative employment is unavailable or out of opportunity. The economic elements impacting business growth include the following:

1. **Capital:** Capital is among the most crucial factors of production when creating a business. Raising capital investment in feasible projects boosts profits, which aids in speeding up capital formation. The business is also gaining traction thanks to the easy access to investments. The existence of capital facilitates an entrepreneur's ability to combine land, machinery, and raw materials to produce goods. For this reason, capital is regarded as a lubricant in the production process.
2. **Labor:** Easy access to the appropriate type of employees also impacts business. The emergence and expansion of a business are shaped more by the quality of work than by the amount. Many less developed nations are labor-intensive because of their large and even increasing populations. However, the possible advantages of inexpensive labor are counterbalanced by the negative consequences of job immobility. The issue of job immobility can be addressed by providing infrastructure, such as effective transportation.
3. **Raw material:** The requirement for raw materials for the onset of industrial activity and its influence on the formation of entrepreneurship does not require emphasis. Without raw materials, it is impossible to start a business or develop an entrepreneur. It is one of the primary elements necessary for production. A lack of raw materials can adversely impact the business climate.
4. **Market:** The market's role is crucial for business development. In the current competitive landscape, no business owner can imagine thriving without up-to-date marketing insights and diverse marketing tactics. The reality is that market potential serves as a significant determinant of the possible advantages of entrepreneurial endeavors. The market's size and makeup influence businesses in unique ways.
5. **Infrastructure***:* Well-developed channels of communication and transportation are essential for the growth of businesses. This not only aids in broadening the market but also widens business prospects. Besides the aforementioned factors, entities like trade and business organizations, business schools, libraries, and so on, play an important role in fostering and sustaining entrepreneurship within the economy.

Social Factors

Social elements can significantly encourage entrepreneurship. Indeed, it was a highly supportive community that contributed to the immense success of the Industrial Revolution in Europe. The social context in which individuals are raised influences their fundamental beliefs, values, and norms. The primary elements of the social context are:

1. **Family Background:** This element encompasses the size of the family, the type of family, and the financial condition of the family. The family's professional and social standing influences mobility and serves as a foundation for entrepreneurship among its members. Research has shown that the sons of business owners possess 2-3 times as many companies compared to those whose parents are not business owners. The entrepreneur acquires significant experience via part-time education and apprenticeships that occur during employment in the family enterprise.
2. **Education:** Education enables the comprehension of the external world and equips individuals with essential knowledge and skills to tackle daily challenges. The education system significantly contributes to instilling entrepreneurial values within each community. While education is not a critical factor for becoming an entrepreneur, it is believed that an educated individual is more adept at grasping the intricacies of business compared to someone who is illiterate.
3. **The attitude of the society:** they pertain to the perspective of the community regarding commerce. In less developed nations, monetary incentives tend to be less appealing, and individuals are not driven by financial gain. Numerous avenues exist for individuals to attain social distinction via non-economic pursuits. This explains why individuals possessing organizational abilities are not drawn to enterprise. They apply their skills for purposes unrelated to finance. Conversely, some societies promote creativity and novelty, thereby welcoming the endeavors of entrepreneurs for profit, recognition, and social standing.

   It is claimed that in 19th century Russia, the elite had a disdain for entrepreneurs. For these individuals, agriculture represented a life of prosperity. They thought that land was God's domain and that the fruits of the land were merely a blessing from God. Throughout this era, Russian folktales, sayings, and music conveyed the idea that becoming wealthy through commerce was inappropriate. Likewise, the rigid commitment to non-violence in Jainism motivated its adherents to participate in trade and abandon farming.
4. **Political support:** Entrepreneurs flourish in a social setting where managers and decision-makers receive ample support. A tranquil social environment and a supportive tax policy motivate individuals to launch a business. Tax incentives and beneficial loans presented by the government inspire entrepreneurs to initiate new ventures or grow current ones. Conversely, communities marked by legal issues like crime, riots, etc. seldom experience business activity. These occurrences can easily be seen in various states of India.

Psychological Factors

Numerous theorists in entrepreneurship have proposed theories that mainly concentrate on psychological elements. They include the following:

1. **Motives:** Motives drive individuals to take action. The expansion of a business

necessitates the presence of appropriate motives, including the pursuit of profit, the acquisition of prestige, and the attainment of social status. When the motives are compelling, ambitious and skilled individuals will take risks and innovate. Cole asserts that entrepreneurs pursue power, prestige, security, and service to society in addition to financial wealth. Stepanek places particular emphasis on non-monetary elements like independence, the self-worth of individuals, power, and societal respect. Rostow examined the transitional changes within entrepreneurial families. He posits that the first generation strives for wealth, the second aims for prestige, and the third aspires to art and beauty.

2. **Need Achievement:** David McClelland offered the maximum essential mental theories approximately entrepreneurship withinside the early 1960s. According to McClelland, the want to prevail is a social power to prevail that characterizes a success entrepreneurs, specially whilst bolstered via way of means of cultural factors. He determined that positive humans, specially people who have become entrepreneurs, had this quality. Also, a few societies generally tend to provide greater humans with a high "want for achievement" than different societies. McClelland attributed this to sociological factors. Differences among societies and people supposed that the "want for achievement" became more in a few societies and much less in others.
3. **Respect for status:** Everett Hagen also highlighted the mental effects of social transformation and described the psychological foundations of entrepreneurship. At a certain time, Hagen claims, numerous social groups significantly lose their standing. Hagen ascribed the diminished regard for group standing to the emergence of entrepreneurship. He contends that the primary factor that eventually results in entrepreneurial behavior is the loss of group standing. He proposed four categories of events that might result in status removal:
    - Group can be overthrown by force
    - Its precious symbols can be underestimated
    - It may lead to a scenario where the situation becomes contentious
    - This stance might not be embraced in the migration industry of the new community

Entrepreneurial Development Programme

The Entrepreneurship Development Program (EDP) is a program that aims to help a person strengthen his entrepreneurial motivation and acquire the necessary skills and abilities so that he can function effectively as an entrepreneur. In other words, EDP concentrates mainly on fostering and promoting entrepreneurial abilities and the growth of a person into a successful entrepreneur. The business development program is made up of three general phases:

A. Initial Phase or Pre-Training Phase: This phase includes:

- Tasks and arrangements required to initiate the training program;
- Recognition and choice of possible entrepreneurs and initial encouragement.

- To find an appropriate company for a prospective entrepreneur.

B. Training or development phase

In this stage, a training program is implemented to improve the motivation and skills of the participants. The objective of this phase is to bring about the necessary changes in the behavior of the trainees. Trainers must assess the degree and level to which the trainees have progressed in business.

C. Post-training or follow-up phase

This phase assesses the extent to which the program objectives have been accomplished. Monitoring and subsequent steps will identify the deficiencies from previous phases and propose recommendations for future policy development. Support for infrastructure, guidance, and help in establishing a new enterprise and enhancing current operations may also be taken into account at this point. The subsequent general steps in the monitoring and follow-up process:

- Establish and handle an individual file for every trainee
- A historical chart that illustrates the personal information of each entrepreneur and the work they conducted
- Contact each contractor and provide the requested information in a timely manner
- We regularly visit each entrepreneur to get the necessary guidance and assistance

Women Entrepreneurship

A women's company is defined as a business that is owned and overseen by women, possessing a financial stake of at least 51% of the capital and offering women a minimum of 51% of the employment generated by the business. Sectors encouraged by women entrepreneurs in India typically include Agarbatti Making, Papadi Making, Embroidery, Handicrafts for Export, Batik Painting, Garment Making, Food Service, Restaurants, Snack Bars, Sweet Stalls, Soft Drink Stalls, and so on., Day Centres, Driving Lessons, Typewriter/ Express jobs, Florists and Dry-cleaning driving, making pickles, maintaining women's lodges, etc.

Women constitute approximately fifty percent of the world's overall population. For many centuries, women were regarded primarily as homemakers. They restricted themselves to the confines of their homes and managed domestic tasks. However, now they have stepped beyond these limitations to engage in various activities. Women are afforded equal treatment in educational professions, healthcare facilities, and management sectors. Global data supports that women have achieved considerable success in numerous areas of endeavor, including academia, politics, administration, and social work, among others. Now they have also started immersing themselves in the industry and successfully running their own businesses. In western countries, women entrepreneurs engage in any activity they want, such as electronics, design, house building.

But Indian women mostly take cheap and low value products like food (pickles, papa, meals) for embroidery, handicrafts, lunches etc. They do not possess talent and resources, family support, and a nurturing environment. Embrace challenges. Consequently, the growth of women entrepreneurs in the nation is exceptional from the perspective of economic development.

Development of Women Entrepreneurs

Entrepreneurship has historically been a male-centric field, but the passage of time has altered that landscape and introduced women as notable and inspirational entrepreneurs today. In nearly all developed nations around the globe, women have entered the business world on equal footing with men.

The typical earnings of women are less than those of men in the majority of countries. In terms of occupations, women are restricted to areas such as education, clerical work, nursing, and healthcare. It is uncommon for them to begin careers in sectors like engineering, business, and similar fields. Despite women constituting nearly half of the global population, their involvement in paid employment remains quite limited. There is a significant number of qualified women who are unable to join the workforce due to socio-cultural factors.

In recent years, women have appeared in various fields of life and successfully compete with men despite social, psychological and economic obstacles. This was possible due to education, political awakening, legal guarantees, urbanization, social reforms, etc. Today, some women in India have excelled in many unusual fields. They are leading consultants, distinguished scientists and mathematicians, famous doctors, interior designers, artists, educators, administrators and managers, and entrepreneurs. They debunked old theories and assumptions about women. The very fact that the world's largest democracy is led by a woman greatly complements the skills and quality of women.

Women Entrepreneurs

The inclusion of women in the business world is a relatively recent occurrence. The businesses established by them tend to be mostly on a small scale. Women entrepreneurs refer to women or groups of women who initiate, organize, and oversee a business. Thus, a female entrepreneur embodies a self-assured, imaginative, and inventive woman who seeks financial autonomy while simultaneously generating employment opportunities for others..

Definition of Women Entrepreneurs

"An organization that is owned and governed by a woman possessing at least 51% of the financial stake and providing at least 51% of the jobs created to women" -By Government of India.

"Women who innovate initiate or adopt business actively are called women entrepreneurs." -J. Schumpeter.

"Women entrepreneurship is based on women participation in equity and employment of a business enterprise." -Ruhani J. Alice

Women entrepreneurs can be broadly categorized into five categories

1. **Affluent entrepreneurs**: These are the daughters and wives of wealthy businessmen. These women have financial help and the necessary resources to initiate a new venture and assume business risks.
2. **Pull factors**: These are educated women living in urban areas with or without work experience who take the danger associated with an emerging enterprise with the assistance of financial institutions and commercial banks. These women take the new business as a challenge to be financially independent.
3. **Push factors:** These women do some kind of business to face financial difficulties. Usually, widows and single women manage an existing family business or start a new business due to difficult family situations.
4. **Rural entrepreneurs**: These women belong to the country and choose the business according to their means and abilities. The applied business requires little investment, minimal risk and does not require special skills.
5. **Self-employed entrepreneurs:** They are uneducated women who are below the poverty line. They select a small and manageable business that is easy to oversee and sufficient to sustain the family.

Main Entrepreneurial Traits of Women Entrepreneurs:

Following are main entrepreneurial traits of women entrepreneurs:

1. **Imagination:** It pertains to a creative method or unique concepts within a competitive marketplace. A well-planned approach is necessary to examine the existing situation and identify business opportunities. It also means that women entrepreneurs deal with experienced people and contract with the right organizations that offer support and services.
2. **Attribute to work hard:** Female entrepreneurs have enough work capacity. Imaginative ideas should be given enough space. Building a business takes hard work.
3. **Persistence:** Women entrepreneurs must have the intention to make their dreams come true. They have to turn a dream into a business. Research shows that successful women have worked hard. They continued to receive loans from financial institutions and other contributions. They survived adverse conditions and adversity.
4. **Ability and desire to take risk:** Readiness denotes the eagerness to embrace risks and the capability to strategize, anticipate, evaluate and calculate. Profit is the reward for risk. Women entrepreneurs take risks, but the risk is well calculated. It offers challenges where the chances of survival and failure are equal.

Financial Assistance

Women experience challenges when trying to obtain loans from banks and financial institutions to launch a business. To overcome this difficulty, financial institutions and banks have special systems where financing for women entrepreneurs is available on favorable terms. Two such important schemes are the Small Industries Development Bank of India's Mahila Udyog Nidhi (MUN) scheme and the State Bank of India's Stree Shakti package.

1. **Mahila Udyog Nidhi:** This is a special scheme that aims to bridge the capital gap when women with insufficient resources dare to set up their own units. The system is a refinancing system for state-owned financial companies and commercial banks. Funds allocated in the system are treated as equity when calculating the debt-to-equity ratio to meet the needs of lending institutions. The scheme offers a soft loan of up to 25 percent of the project cost up to Rs. 2.5 lakh per project at zero interest, but only one percent service fee is charged to cover the cost of the lending institutions. To get this help, the cost of the project should not exceed Rs. 10 lakhs. Women entrepreneurs who start new projects, expand, modernize, technologically update and diversify an existing unit, as well as re-establish viable sick manufacturing or service sector units can apply for assistance.
2. **Stree Shakti Package:** This is a unique scheme of State Bank of India to support women entrepreneurship. Companies where women own more than 50 percent of the share capital can apply for it. Stree shakti package offers following benefits.
   - The margin shall be reduced by 5 percent according to such type of industrial units
   - If the loan amount exceeds Rs. 2 lakh interest was calculated at 0.5 per cent
   - In small sector units, no collateral is required for a loan up to Rs. 5 lakhs

The government has launched several programs to promote women's entrepreneurship, and the success of women's entrepreneurship education programs is better.

Case-A Successful Dairy Farm Intervention

Manasi Ksheerasagaram unit is a thriving entrepreneurial initiative established by five women: Moli Varghese, Shylaja Raju, Rajini, Padmini, and Lalitha Surendran. The unit operates under Kudumbashree CDS of Melur Gramapanchayat in Thrissur district. This venture serves as the primary source of income for the women involved. All the members derive significant satisfaction from their agricultural activities and the earnings that result from them. The unit maintains a total of 14 cows, which include both heifers and lactating cows. The products they market from the unit consist of milk, ghee, curd, and dung.

The Manasi Ksheerasagaram unit was established in 2019 with the subsidized loan obtained from Kudumbashree, the state's poverty alleviation initiative that is executing the DAY-NRLM. At the outset, the women faced challenges in securing an income for their livelihoods. During that time, Kudumbashree provided funding assistance to launch a farm. The women seized this opportunity and embarked on the venture with courage and optimism.

Now they are receiving feed at a subsidized price from the milk society. Additionally, they are engaged in fodder cultivation. Kudumbashree assisted them in constructing a cattle shed to buy cows to establish a fodder cultivation unit through a subsidized loan.

Now the mentors are quite satisfied with their lives. The unit fulfills their daily financial requirements, including their children's education, thus acting as the financial foundation of their families. The women state that such initiatives from Kudumbashree empower women by providing them with both social and financial status. The unit generates a monthly income of Rs. 2,06,500/- and after subtracting the monthly expenses for feed and other costs amounting to Rs. 22,250/-, they achieve a net profit of Rs. 1,84,250/-.

Kudumbashree proves to be the genuine breadwinner for the aforementioned women and consequently for their families. A significant amount of profit earned by this group of women serves as a role model to society, and many similar groups have been established in the vicinity, thus paving the way for the development of the local community.

Source: Inspirational Stories of Aatmanirbhar Rural Women (https://agriculture.vikaspedia.in/viewcontent/agriculture/women-and-agriculture/a-successful-dairy-farm-intervention?lgn=en)

Self Help Groups -What are SHGs?

Self Help Groups (SHG) are casual collectives of individuals who unite to seek methods to enhance their living standards. These groups are typically self-managed and centered on peer support. Individuals with comparable economic and social backgrounds often engage with the assistance of an NGO or government organization and strive to address their issues and better their living conditions.

Functions of Self-Help Groups

- They aim to enhance the abilities of the impoverished and marginalized communities for job opportunities and income-producing activities.
- They provide unsecured loans to groups of individuals who typically struggle to obtain loans from banking institutions.
- They further mediate disputes through cooperative dialogue and shared leadership.
- They serve as a vital provider of microfinance services for the underprivileged.
- They function as a conduit through which formal banking services are accessible to the poor, particularly in rural regions.
- They additionally promote the practice of saving among low-income individuals.

Need for Self Help Groups

- One of the primary causes of rural poverty is the restricted or nonexistent access to credit and financial services.
- The Rangarajan Committee Report pointed out four key factors for the absence of financial inclusion in India. They are:
  - Lack of ability to provide collateral security
  - Low credit absorption capability
  - Inadequate reach of institutions
  - Fragile community network

It is recognized that a key element of rural credit unions is the existence of robust community networks in Indian villages. Self-Help Groups (SHGs) are crucial for delivering credit to the impoverished, which is essential for reducing poverty. They also significantly contribute to the empowerment of women, as SHGs assist women from underprivileged backgrounds in developing social capital. Achieving financial independence as a self-employed individual further enhances other development aspects like education, health services, and improved family planning.

Advantages of Self-Help Groups

- Financial Inclusion – SHGs promote banks to provide loans to the impoverished and marginalized groups in society to guarantee returns.
- Voice to marginalized – Self-Help Groups (SHGs) have provided a platform for the otherwise underrepresented and voiceless portions of society.
- Social Integrity – SHGs contribute to the resolution of various social issues such as dowry, alcoholism, early marriages, etc.
- Gender Equality – By empowering women, SHGs assist in guiding the nation towards genuine gender equality.

- Pressure Groups – SHGs function as pressure groups through which the government can be urged to address significant issues.
- Enhancing the efficiency of government schemes – SHGs aid in the implementation and optimization of government systems. They also help to mitigate corruption through social audits.
- Alternate source of livelihood/employment – SHGs assist individuals in earning a livelihood by offering vocational training and enhancing their existing livelihoods by supplying tools, etc. They also contribute to decreasing reliance on agriculture.
- Impact on healthcare and housing – The financial inclusion facilitated by SHGs has resulted in improved family planning, decreased child mortality, better maternal health, and has empowered people to combat disease more effectively through enhanced nutrition, health, and housing.
- Banking knowledge – SHGs motivate individuals to save and enhance banking knowledge in rural regions.

STEPS IN SETTING UP AN ENTERPRISE

1. **Search for business ideas:** The initial phase in the entrepreneurial framework is to seek out business concepts. An inspiration can originate from various sources, including a family member or friend's success narrative, the need for specific goods, attending fairs and exhibitions, examining project outlines, market feasibility assessments, consultations with governmental bodies, and so on.
2. **Idea Processing:** Once business concepts are identified, these concepts are evaluated and examined to determine their technical feasibility and market viability. Technical feasibility is evaluated based on the accessibility of required technology, machinery and equipment, skilled workforce, and raw materials. Business profitability pertains to the profitability analysis based on the investigation of potential demand, anticipated sales volume, selling price, production expenses, break-even point, etc.
3. **Idea selection:** After evaluating the different aspects of each concept, a selection matrix is created that examines the diverse data gathered to choose the most feasible idea. Following the selection of a business concept, the results are compiled into a document referred to as a "project report" or "business plan," which outlines the methods for effectively establishing and managing the business. This document is beneficial for securing financing, permits, and so forth.
4. **Choosing a business location:** To attain success, a business owner must select a customer-oriented site and adhere to the regulations of the nation's commercial area.
5. **Business financing:** If an entrepreneur lacks sufficient funds to initiate a business, alternatives exist in the shape of government-supported schemes,

bank financing, venture capital, research grants, etc. Numerous programs are accessible to assist various categories of entrepreneurs in launching their ventures.

6. **Determine the legal structure of the company:** The subsequent step involves the entrepreneur determining which ownership structure is most suitable - for instance, sole proprietorship, company, limited liability company, corporation, non-profit organization, or cooperative..
7. **Obtaining a tax registration number:** An entrepreneur is required to register a business with a "business name" and register with a government body to acquire a tax registration number, employee pension fund, workers' compensation, unemployment and disability insurance, etc.
8. **Obtaining Business Licenses and Permits:** As per the regulations of the nation, the business owner is required to acquire the essential licenses and permits from the Central Government/State Government in order to operate the business.
9. **Selection of personnel:** Personnel is the company's most valuable asset. The contractor must decide on the following issues:
    - The quantity of staff needed for corporate administration, technical, and additional responsibilities
    - Qualification and experience of the personnel necessary to carry out the work effectively
    - The process and techniques to choose the most suitable candidates
    - Staff performance assessment system and standards
    - Principles and methods of personnel compensation
    - Services to guarantee worker safety, health, and well-being
10. **Establishing a business:** This is the concluding phase where the entrepreneur collects all the required resources and sets up the business.

Few entrepreneurial development models

1. **Entrepreneurial development models:** The frameworks proposed for fostering entrepreneurship can be categorized into the following groups:

   a) The Psychological Models

   - McClelland (1961) gave an important lead in identifying the factors affecting entrepreneurship. In his model, he considers achievement motivation, which was previously associated with parenting practices, as even more important. He changed his previous understanding of the achievement motive as an inherent factor, and now the change in motivation is seen primarily as a result of the ideological awakening of the latent need for achievement in adults. Having identified achievement orientation as a key variable in entrepreneurial development, McClelland proposes a motivational training program as a policy measure to make them truly ready and eager to exploit the new opportunities they offer.

- Everett Hagen's theory of social change emphasizes "creative personality" as a causal link in entrepreneurial behavior and "status withdrawal" as a determinant of "creative personality." Hagen explains in detail the causal sequence of entrepreneurial behavior. But his business model does not provide political variables for business development. The regression of status would occur in the natural evolutionary process of society and under no circumstances.
- John Kunkel (1965) addressed the supply side of entrepreneurship by proposing a behavioral model. His model suggests that entrepreneurial behavior is a function of the surrounding social structure, both past and present, and can in fact be influenced by manipulated economic and social incentives. Thus, the model of Kunkel is based on experimental psychology, which identifies sociological variables as determinants of the provision of an entrepreneur.

b) The Sociological Models

- Frank W Young's entrepreneurial theory is a transformative theory rooted in the participation of reactive subgroups within society. In general societal contexts, the re-engagement of a lower-status subgroup stimulates entrepreneurial activities when that group possesses superior institutional resources compared to their peers at an equivalent societal level. Young's model of entrepreneurship suggests that the establishment of societal support institutions serves as the key determinants of entrepreneurship.

c) The Integrated Models: T.V. Rao (1975) proposed the subsequent framework for entrepreneurial development:

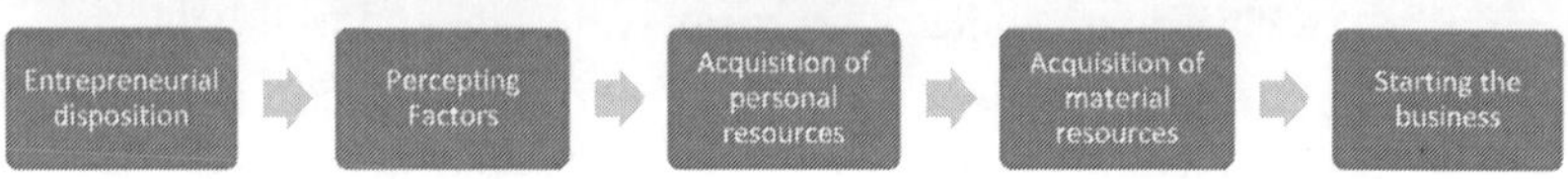

In 'Entrepreneurial Disposition' he includes the following factors:

- **Need for Motive**: The dynamic which for the prospective entrepreneur has the greatest possibility of achieving the goals if one performs those activities.
- **Long Term Involvement**: The goals either at thinking level or at activity level in the entrepreneurial endeavor that is regarded as a goal to be achieved.
- **Resources**: Personal, social and material resources which he thinks are related to entry and success in the area of entrepreneurial activity.
- **Socio-Political System**: A system perceived as suitable for establishment and development of his enterprise.

All these elements combine additively in character, and their optimal presence leads to the starting point of entrepreneurship, which leads to the acquisition of material resources and the establishment of a business.

2. **Five stages for promoting small entrepreneurship:** B.S. Venkata Rao (1975) has outlined five phases for encouraging small business ownership namely,

   a) **Stimulation:** This stage involves the development of an industrial environment, orientation focusing on the importance of small-scale industry, extensive promotion of the industrial development initiative, the creation of specialized systems, and the formation of supportive institutions. This step is necessary to create interest and awareness in backward areas for industrial activities.

   b) **Identification**: This step is necessary to identify potential entrepreneurs. The entrepreneurial force of the future can be identified among rural artisans, factory workers, individuals with formal engineering and technology education, as well as business and management graduates.

   c) **Development:** This phase would include the organization of motivational and management training programs as well as technological advisors, planning of the funded project, location, etc.

   d) **Promotion:** These can be job support institutions that offer development opportunities, services and incentives.

   e) **Follow-up:** At this point, government policies and programs are evaluated and new methods of enhancing their effectiveness are pursued.

3. **Cycle for the promotion and development of entrepreneurship:** M.M.P. Akhori (1977) proposed the subsequent cycle for the advancement and enhancement of entrepreneurship:

   a) **Stimulatory activity:** The stimulatory activity refers to all such activities that stimulate entrepreneurship in any society.

   b) **Support activity:** Support activities mean all such activities that assist the businessperson in founding and overseeing the organization, as well as maintenance activities that promote the continuation of the company's efficient and profitable operations.

   c) **Sustaining activities**

4. **Stages of economic growth:** K.L. Sharma (1978) describing four phases of economic development, specifically,

   a) Engagement in manufacturing

   b) Growth of business

   c) Assumed stability in business, and

   d) Dedication to ongoing growth.

   The entrepreneurial growth is governed by three sets of entrepreneurial determinants, namely,

   a) Need for achievement motivation,

b) Socioeconomic background of entrepreneur (family occupation, caste, religion and region), and

c) Political milieu.

The interaction between these stages of business growth and the factors determining business growth does not happen in a vacuum, but in the presence of prerequisites for business, such as material resources, the vision of the entrepreneur and self-confidence in this field, and the general. infrastructure available to the contractor. According to this model and empirical analysis, K.L. Sharma suggests that family occupation is more effective at industry entry and political factors have a positive effect on unit expansion and these factors are more important in the third stage, i.e., experiences firm stability.

5. **Models of Corporate entrepreneurship:** Robert C and Michael J had discussed the four models of corporate entrepreneurship. It includes Opportunists, Enabler, Advocate and Producer.

   a) **The Opportunist:** However, the opportunistic model only works well if it relies on corporate cultures open to experimentation, with diverse social networks behind formal hierarchies.

   b) **The Enabler:** Unlike the opportunistic model of distributed ownership and ad hoc resource allocation, the enablement model has its own resources. "The early stages of new business planning are supported, encouraged and often strategically guided by promising concepts that have promised serious management attention." But the enabling model is not just about allocating capital to business ventures. It is also about personal development and leadership responsibility.

   c) **The Advocate:** In the third model, activist (with targeted ownership and ad hoc resource allocation), the company gives ownership of the organization to a firm-level group appointed to create new projects, but intentionally gives the group only a modest budget. Evangelists and innovation experts act as evangelists and innovation experts, promoting business entrepreneurship with business entities having to demonstrate their commitment to new business development by paying most of the bills, as the authors note.

   d) **The Producer:** The fourth model, with concentrated producer ownership and dedicated resources, aims to protect new projects from turf battles, encourage cross-unit collaboration, create potentially disruptive companies, and create opportunities for executives to pursue careers outside their business units. Wolcott and Lippitz explain.

   Various social scientists have extensively acknowledged entrepreneurship as a significant factor in economic advancement. The idea of entrepreneurship is adorned with several characteristics such as "innovation", "organizational building capacity", "filling function", "input completion", etc. Individuals with these characteristics are required to initiate and maintain the industrialization process. Efforts must therefore be focused on finding and developing such entrepreneurial talent.

Social Entrepreneurship

Social Entrepreneurs are people who are prepared to generate beneficial transformations in society through their creative concepts and endeavors.

- They manage their business or organization to achieve their goals by helping society.
- Their motto when they start a business is mainly to help society and they have no great intention of making personal profit.
- Their success is not always measured by profit alone. Success is also a small change in society from their efforts.

Social entrepreneurship is also referred to as altruistic entrepreneurship –which translates to selfless concern for the welfare of others. Numerous environmental and social issues out there and social entrepreneurs identify those problems and come up with innovative ways, and establish or adopt a business model around it.

By presenting user-friendly and adaptable ideas to the local people, social entrepreneurs are generally looked up to as leaders or role models for driving philanthropic projects and bringing a large group of people to believe in their initiative. In the recent past, there have been many social entrepreneurs or philanthropists who have made immense success and proven to the public that not all needs of society are to be solved by the Government. Any person with a vision and the ability to bring change can do it.

Below are a Few Examples of Leading Social Entrepreneurs

- Vinoba Bhave (India) was the leader and founder of the Land Gift Movement. He led to the redistribution of around 7,000,000 acres of land that later on helped the landless and untouchables of India.
- Dr. Maria Montessori (Italy) – The Montessori approach to early childhood education was developed by her.
- Elon Musk, Tesla Motors, SolarCity, and SpaceX caused Musk to become an explorer of social entrepreneurship in the modern era, as he pursued to create solutions that are accessible to renewable energy and push the bounds on space exploration for the human race.

  Social entrepreneurship has similarities and distinctions when compared to the standard definition of entrepreneurship.

Importance of Social Entrepreneurs

- **Social Impact and Inspiration:** the origination of a social enterprise is an existing gap or problem that an individual identifies and has a solution for. Social entrepreneurs' works tackle the existing problem of society; they do not intend to create legacies and are an inspiration to society.

- **Make the World a Better Place:** Social entrepreneurs are obsessed and extremely passionate about the initiative and work towards the goal against all odds, and can go to any extent to see that society is problem-free. It is the social entrepreneurs who can bring drastic changes to culture, business, and economy while they make a living out of it.
- **Generation of Social Capital:** One of the most important values and powerful influences created by Social entrepreneurs is the equation established by social groups through interpersonal relationships, a shared sense of identity, and a shared understanding in their network that helps people to 'get by, accept and get ahead' with changes.

Factors the Strengthen Social Entrepreneurship

- **Project Contributes to the Economy:** Having known social entrepreneurship is for the society or environment, it also means there is or has been a demand for the product or service. Most entrepreneurial initiatives contribute to the economy, by creating job opportunities and wealth. The social enterprise thus established must aim at generating enough wealth that can contribute to society.
- **Responsibility towards Society and Environment:** The primary intention of social enterprise is to identify gaps in the environment and society that are not working efficiently and create a social value out of those. Aimed at bringing in a change and something new to solve a certain problem, these objectives can vary from industries, health services, education, energy-saving, etc. A lot of corporate entities form a small trust to implement these objects in the form of Customer Social Responsibility (CSR) projects.
- **Effective Profit Utilization:** As discussed, earlier social entrepreneurs are not aimed at making personal profits, their profits are often re-invested in the business to achieve the goal, and personal profits are hence supervised and are limited. However, based on the project's status, the entrepreneur can decide how much is to be reinvested to achieve the goal.
- **Efficiently Managed:** Unlike the standard entrepreneurship framework, social entrepreneurship can involve a group of people working together with the same intention of bringing a change to society. Hence, the decision-making, execution of tasks, etc. are all shared and are done with the active participation of all experts at different levels. Here, making use of participation efficiently is the key.

Types of Social Entrepreneurship

- Non-profit organization
- Co-operative societies
- Social enterprise
- Community project
- Special purpose business

Characteristics of Social Entrepreneurship

Social entrepreneurs aim of contributing to the society for the long-term business activities in the society. Their main focus is towards creating social capital without measuring the performance in terms of profit. Some characteristics of social entrepreneurship include the following:

- **Achieve Social Change:** Social entrepreneurship is considered as the means of achieving change in the society. It recognizes the social problems, achieves social change by employing entrepreneurial principles, processes and operations.
- **Initiative Innovative:** There is initiation of social entrepreneurship towards the innovation of new concepts, ideas, and models for solving social problems. The problems like health, education, unemployment, natural calamities, etc. are considered as social problems. There is generation of new methods and techniques for resolving social problems and issues by the social entrepreneurs.
- **Create Value:** The main aim of social entrepreneurship is creating value to the society. It focuses on creating social capital without measuring the performance in terms of profit or return. Entrepreneurs perform such business activities that can create and maintain value of the society.
- **Change Agent:** Social entrepreneurs are the change agent of the society. They support the society through their innovative new concept, ideas, and knowledge. They understand the social needs and fulfill it through creative organization. They measure their success through improvement of social value and standard, not only in terms of profit.
- **Feel Satisfaction:** The application of innovative, practical, sustainable, market-based approaches is emphasized by social entrepreneurship which support to fulfill the needs of under-served populations. In present time, it has become the priority for decision-makers at the most senior levels.

8

Creativity, Innovation and AGRO Entrepreneur

Introduction

Creativity and innovation are planned alternatives that an entrepreneur can use to promote his business. Business is a creative process. Success in business today requires constant innovation. Creating new solutions to problems, and the ability to create new products or services for a changing market, is part of intellectual capital. They also provide a company or business organization with a competitive advantage. Creativity is the basis of creativity. It is a process and skill that can be developed and managed throughout the organization.

Learning Objectives

- Describe the different types of creativity and Innovation
- Explain the process of creativity and phases of innovation

Creativity Definition

Creativity has different meanings. Many researchers have said that creativity is the generation of new ideas that include a creative process or a new solution to a problem and the original resolution of problems. It can also be said that the installation of the innovative solution is easy.

Types of Creativity

In general, the quality of construction determines the type of innovation. Different authors have proposed different types of creativity. Irwin Taylor proposed the following quality hierarchy:

Spontaneous creativity: it's about creativity.

Technological Innovation: involves a significant improvement in a process that increases the level of skill or efficiency.

Creativity: a creative combination of materials or ingredients. For example, an Edison lamp or a telephone.

Creativity: involves the use of more basic ideas such as management principles for psychological principles to create a more effective system for motivating employees.

Creativity: involves the revolutionary new principles of an art or science, such as

Freud's psychological theories, the relative theory of Einstein's or Picasso's cubism ideas. These are just creative types, each one is different. The quality and level of innovation is related to the degree of innovation and accuracy of the innovation.

Creative Process

The creativity process in organisations involves three steps

A. Generation of ideas

Generating ideas in an organization depends on the flow of people and information between the company and its surroundings. They are highly motivated by external relationships. For example, most technological innovations have been made in response to market conditions. If the manager knows the demand for a new product or is not satisfied with the current product, he can find new ways to maintain market share. It generates new ideas on the basis of opinions of external consultants and experts. Generating ideas within a nurturing environment serves as a crucial instrument for executing innovation.

Idea development

Conceptual development depends on the organizational culture and processes within the organization. Organizational characteristics, values and processes can support or hinder the use and development of innovative ideas. Organizational structures that prevent communication between departments often prevent effective people from recognizing a problem.

Implementation

The implementation phase of innovation in organizations includes those that bring solutions or innovations to market published for.

Techniques of creative thinking

The following techniques are useful in generating creative ideas:

Brain storming: This is a group technique, but individuals can use it. The idea behind the idea is that quantity of the ideas will result in the quality and the acquisition of many ideas. Always encourage more ideas. Thinking is a very clear problem that has many possible solutions. For example, how to increase the productivity of the plant, or how to increase the sales of a product. Brainstorming allows members to begin thinking of new solutions to a problem. The functions of thinking.

- There should be a specific purpose for brainstorming.
- Many ideas.
- Provides additional options. It is widely accepted.
- It is not useful to mean that the problem has a unique solution that can be analysed.

B. Attribute analysis

It tries to find ways to satisfy the key features of a product or service. It is good for designing or redesigning a product or service or a specific function. It is good for designing or redesigning a product or service or a specific function. It includes functional analysis.

- Identify the key features of a working demonstrator
- Create different ways to ensure performance
- Evaluate different designs based on criteria such as product design, cost, marketability, and others.

The more specific the object or activity that a person wants to change, the better the results that can be achieved through job listing. It forces you to focus on situations that can be changed. An example of the process:

- List all visible properties of an object or activity such as current size, color, shape, functions, weight, priority, steps in an activity
- Define some of these features can be changed without destroying the original work. of an object or activity
- Presenting different characteristics as separate parts.

Story boarding

It's a creative method of strategic planning and situational awareness that teams use the most. Holds 8-12 people. There are four stages: planning, thinking, managing and communicating. The ideas generated from the brainstorming are arranged in a logical order on the board and form a narrative. This helps to connect all the ideas and fix all the parts. It makes it easy to determine ways to implement solutions.

Lotus blossom

Designed for teams, they are used to provide different and deeper solutions to a problem. It starts with the main idea in a circle surrounded by 8 empty circles. 8 more ideas are generated through brainstorming, each circle is surrounded by 8 more empty boxes and filled in through discussion. This process continues until a satisfactory solution is found.

C. Checklist of questions

An initial answer to numerous inquiries. It is primarily utilized to enhance or alter the product. For instance: How can the product be adjusted to ensure it is more cost-effective? The application of a series of terms, action words, descriptive words, or expressions in lists or charts related to a product or service or its characteristics.

Process of Creative Problem Solving

Stages are involved in the process of creative problem solving:

Preparation: It offers a chance to analyze the issue from every perspective, encompassing a comprehensive grasp of the underlying cause of the issue and the limitations that must be addressed. This ought to be meticulously explored utilizing practical reasoning methods.

Incubation: Reassigning the issue to the conscious mind and permitting it to simmer beneath the surface of awareness. Understanding is frequently required to produce insight to address intricate challenges.

Illumination: Illumination is a form of inspiration, often associated with two distinct effects with light: a sense of power and energy.

Verification: A development or validation of a vision solution. It also applies to concepts and refines basic concepts.

D. Innovation

Innovation pertains to the creation of original ideas and novel activities, which convert the original idea into a new product or operation. Innovation involves not merely engaging in something unprecedented, but rather redoing established concepts in a novel manner. Innovation represents the conversion of inventive ideas into practical uses, whereas creativity encompasses the act of inventiveness. Innovation serves as the distinctive instrument for entrepreneurs. Innovation constitutes an evolutionary journey.

Characteristics of Innovation

- Innovations that demand and lead to change
- Innovations that arise spontaneously or emerge from a well-thought-out plan
- Innovations that lead to radical change or expansion or common of existing products and services. structure /nature of the business
- Innovation is a work-in-progress• Innovation helps simplify products and processes
- Innovation is always testing, testing, testing and the modifier.

The Elements of Innovation

- Innovation serves as the fundamental element of entrepreneurship. The existence of creativity distinguishes it from others. Below are the attributes of innovation as outlined by Fiona Fitzpatrick.
- Challenge - the idea of transformation - "pull"
- Customer focus - value for customers is generated - "push"
- Innovation - produce and exchange ideas - "brain"
- Communication - the circulation of ideas and information - "lifeblood"
- Collaboration - individuals unite to brainstorm - "heart"
- Completion - execution of a new concept - "Muscle"• Ideation - The contemplation of experience enhances understanding – "Stige" •
  Culture – the imaginative space where leadership, diverse groups of individuals, core values, and creative principles coexist. Business

- Context- Innovation arises from engaging with the world.

Types of Innovation

Types of innovations; technical and non-technical

Technical innovation: This is connected to innovation in the methods by which production occurs, along with innovation in the products themselves. It is also referred to as pure creativity. Crafting something entirely distinct from current technology or products. Product-based innovation is described as the creation of goods and services that encompass new advancements and new advancements.

Non-technical innovations: It is referred to as creativity as well. It signifies growth and evolution. The objective is to enhance current industries. It pertains to minor modifications in design, product manufacturing and production, materials, and service delivery that companies implement to sustain their productivity and decrease expenses.

Innovation also can be classified as

Invention: The creation of a new product, service or process

Extension: The development of an existing product, service or process. This is a different operation than currently thought.

Duplication: Copying is not just copying but adding your own creative flair to enhance and enhance the concept to beat the competition.

Syntheses: A blend of pre-existing ideas and components in an innovative framework. It entails utilizing numerous concepts and elements that have already been developed and discovering methods to integrate them into a new initiative.

Sources of Innovation

Unpredicted occurrences: At times, unanticipated success or failure can reveal a significant innovative surprise for the organization. Such occurrences are typically expected and not planned.

Incongruities: This happens when there is a gap or difference between expectations and reality. This opportunity encourages the entrepreneur to be more creative.

Process needs: These needs arise when the situation and environment create demands for innovation to meet process needs.

Industry and market changes: Changes in consumer needs, styles, and technological advances require changes in product quality. These changes provide great opportunities for innovation.

Demographic changes: Urbanization, increasing education, employment, income, etc. provide more opportunities for the creative entrepreneur.

Perceptual change: A change in attitude, emotion and interpretation of a person towards a subject lead to a new process.

Knowledge based concept: At times, unexpected success or failure can uncover a substantial innovative surprise for the organization. Such events are often anticipated and not premeditated.

Different Forms of Innovations

Product Innovation

Product innovation arises from discovering new methods to address a customer's issue by creating new products or modifying the features of an existing product for the advantage of both the customer and the supporting company. Example:

- Automobile: Presenting advanced models of cars and motorcycles
- Agriculture: Presenting improved types of pesticides

Process Innovation

Process innovation entails the introduction of new process technology or the enhancement of current technology to an advanced level to enhance performance. Innovation boosts the bottom line, cuts costs, heightens efficiency, and enhances productivity.

Business innovation

Business innovation encompasses several critical subjects, such as the creation of novel business models, innovation within organizations, the application of technology and communication in business, fresh management practices, environmental efficiency, emerging forms of stakeholder engagement, transportation, and finance. Innovative business models, contemporary management frameworks, new methods for managing value chains, alternative communication strategies, approaches to thinking and knowledge management, different varieties of strategic alliances, and unique sales and customer service methods.

Organizational Innovation

More efficient tasks associated with novel management practices, demonstrating recognition that innovative methods of overseeing work in fields like personnel management by enabling the workforce, bringing in new personnel or effective approaches to engage all employees in transforming work management into team resources for creativity, knowledge management, value chain management, customer engagement, distribution, finance, and production can enhance competition. Organizational innovation encompasses new business models.

Technology Innovation

Technological advancement is the outcome of the research and development of technology that pertains to particular products and services.

Marketing Innovation

Innovative distribution methods and customer service are significant elements. They assist a company in generating new value-added services, penetrating new markets, and developing new segments of the market, along with new distribution techniques and enhanced features of customer service and integration with the customer. Marketing communications can additionally be more impactful with imaginative strategies.

Strategy Innovation

It includes business renewal strategy, new business growth strategies, better competitive strategies. Challenging existing industry trends to generate customer value in order to fulfill the requirements of new customers, enhancing value, and establishing new markets and new customer segments for the supporting company.

Phases of Innovation

Innovation has two phases

- Design phase
- Implementation phase

Design phase

The design phase involves approaching tasks with a different mindset and exercising creativity, thinking innovatively, and exploring various options. The administration of the design phase ought to be defined by straightforward management and oversight by the professional in a given circumstance, transparent communication, the exchange of information, numerous discussions, and so on.

Implementation phase

It requires much planning and careful coordination and control and evaluation of progress.

Risk, Uncertainty and Innovations

Creative entrepreneurs embrace unpredictability. Unpredictability is a circumstance where the entrepreneur recognizes the issue but lacks complete information regarding the potential solutions and the outcomes that may arise from each. The various categories of unpredictability are as follows:

Structural uncertainty: This is connected to specific structural conditions and limitations placed on new companies due to the arrival of fresh investments. Technological volatility serves as the prime illustration of structural instability.

Strategic uncertainty: In situations where complete information is lacking about the extent of competition, the quantity and identities of rivals, the categories of products and practices utilized by competitors, and governmental regulations concerning the new industry. New investors in developing sectors are unaware of these crucial variables, and this kind of uncertainty poses challenges for them.

Resources uncertainty: It pertains to the accessibility of resources in the company's target markets. New enterprises frequently encounter challenges in their efforts to secure financing from the financial markets since the sources of funding are not well-acquainted with the risk and return associated with new enterprises. Professional executives also encounter economic and financial unpredictability when they enter organizations within emerging sectors. As the sector grows, acquiring raw materials and components becomes increasingly challenging.

Customer uncertainty: It pertains to the production or consumer marketplace. Typically, there is insufficient information regarding the desires and demands of the consumer, income brackets, demographic traits, psychological traits, and consumer perceptions, etc., and fresh investments in growing industries will solidify this data. They additionally encounter consumer pushback concerning product pricing and rates.

Innovation signifies the initiation of an unforeseen journey. Insecurity has various origins. It might not be apparent what method is necessary to accomplish product innovation.

Creativity, Innovation and Entrepreneurs

Entrepreneurship = Creativity + Innovation

Creativity – thinking new ideas

Innovation- Doing new things

Entrepreneurship -Creating value for the new idea in the market

Entrepreneurship arises from a process that employs creativity and innovation to access the market. Entrepreneurship encompasses far more than merely initiating a new business. It pertains to all the efforts that empower individuals to comprehend the dynamics of personal life, the evolution of concepts, and assume responsibility and personal initiative. In a broader sense, entrepreneurship aids young individuals in acquiring new skills and experiences that can be utilized to address many of the world's issues. Thus, entrepreneurship stands as the primary focal point and the capability to ignite activity and generate value through creativity and innovation.

Growth and advancement cannot persist without innovation (often related to a product or service or its promotion) and other innovations, as well as "simple" companies. It serves as the motivating force for ongoing growth and development.

Successful entrepreneurs require a level of excellence that arises from a blend of imaginative thinking and the capacity for execution. An entrepreneur's innovation might entail a novel product or process that alters the current order, or the entrepreneur possesses distinctive insight into the process or results of external change. Entrepreneurship serves as a mechanism that fosters creativity and innovation. Innovation generates new demand, while entrepreneurship introduces innovation to the marketplace. Innovation represents the effective cultivation of competitive advantage, making it essential to entrepreneurship.

Case-An Innovation that Generates Revenue for Rural Women

Mr. K. Vivekanandan from Coimbatore (Tamil Nadu) invested Rs. 8 lakh and developed a 3 HP pin pulverizer for grinding chili and coriander. "The machine serves as a perfect source of income for rural women who wish to enhance their household earnings," states Mr. Vivekanandan.

Most of the current devices for grinding chilli and coriander necessitate substantial installation expenses and consume a significant amount of energy, rendering them impractical for rural regions, where the electricity supply might also be unreliable.

Challenges faced

Mr. Vivekanandan believed he addressed 90 percent of the grinding issue when he created the machine and produced close to 100 units. However, he was disappointed to discover that only 20 units sold. A few of the purchasers returned the machine because the chilli and coriander did not go through the filter screen and generated excessive dust during the grinding process.

Work came to a standstill, and did not resume for almost a year.

Mr. Vivekanandan learned about Villgro (an organization that assists rural entrepreneurs) and he reached out to them for support. The team at Villgro utilized various resources to address this issue. Technical expertise initially helped Mr. Vivekanandan create a 1 HP, single-phase machine, as the machine was unable to operate at a speed of 3 HP initially. (In rural regions, a one-HP, single-phase machine is preferred due to voltage fluctuations).

After multiple attempts, they discovered that the issue of chilli and coriander getting stuck on the screen was not caused by their high fiber content, but rather by the rotor's speed. Consequently, the machine's weight was decreased, and adjustments were made to its wall thickness, dimensions, and the diameter of both the stator and rotor to accommodate rural applications.

Cost

Mr. Vivekanandan reduced the price of the machine to meet rural requirements by concentrating on the types and quantities of materials utilized. Each unit is sold for Rs. 11,500 (including motor).

For more information readers can contact:

Mr. K. Vivekanandan, M/s Vivega Engineering Works, New No: 116-118, Sathy Road, R. K. Puram, Ganapathy, Coimbatore - 641 006. Mobile No: 94437-21341.

Source: A innovation that generates revenue for women, Science & Technology column, The Hindu

Business incubation

Incubators are institutions that support entrepreneurs in developing their business, especially in the early stages. These are organizations designed to accelerate the growth and success of startups and early-stage companies. It is usually done by companies with experience from business and technology.

Incubation support includes providing input and technical advice, initial growth funding, partnerships and alliances, co-working spaces, labs, consulting and advisory support. These are good ways to attract capital from angel investors, government agencies, economic development coalitions, corporate bodies and other investors.

Most incubators have potential funds to invest in growth startups or are affiliated with potential funding sources. They provide access to compliance services from professionals such as accountants and lawyers. It goes without saying that the staff and other entrepreneurs at the Growth Center offer great advice and networking support at the Incubation Center.

Incubators serve as a crucial early stage component of the startup ecosystem. They will act as a driver for local and national economic growth. Various types of incubators exist: universities; non-profit development organizations; investment development for social good; cosmetics firms, and the aforementioned categories.

Incubators have different strategies. Some of them are in the physical space, strengthening the relationship between business incubators, and others are in the room of the entrepreneur. others work closely. Incubators are sometimes referred to as accelerators, often aimed at more developed startups. Venture capitalists also differ from research and technology platforms in their commitment to support startups and early stage companies. Research and technology platforms are an important topic in organizations ranging from public institutions, companies, university laboratories to very small companies. Unlike business incubators, it is difficult to offer business support services. However, it is safe to say that business support services are the hallmark of incubator businesses.

9

Business Opportunities and Business Plan for Startup

STARTUP India is an innovative initiative started to help people who want to start their own business. These individuals are reflective and resilient. Therefore, the government provides assistance for them to realize their concepts and expand. The Government of India introduced the 'Startup India' initiative on January 16, 2016, in New Delhi.

What is a start up?

According to the government notification, the journal number will replace the G.S.R. 364(E) dated 11 April 2018 as amended by Notification No. G.S.R. 34 (E) on 16 January 2019 by DPIIT (Department for Promotion of Industry and Internal Trade).

Definitions

In this announcement: An entity shall be measured as a Startup:

- Any entity for a period of ten years from the date of incorporation/registration, if registered as a private limited company (as defined in the Companies Act, 2013) or listed as a partnership (registered under Section 59 of the Partnership Act, 1932) or a limited liability partnership (under the Limited Liability Partnership Act, 2008) in India.
- The revenue of the entity has not exceeded one hundred crore rupees for any of the financial years since registration/registration.
- The business unit works to innovate, develop or improve products, processes or services, if there is a scalable business model with high potential for performance or value creation. But the enterprise created by revolution, by revitalization of the existing business, is not called a "start-up".

Definition A business ceases to operate after ten years from the date of incorporation/registration or if its income exceeds one hundred crore rupees for any previous year.

A startup, also known as a startup, is a company or project launched by an entrepreneur to find, develop and validate a business model that can be scaled. While entrepreneurship refers to all new businesses, including freelancers and unregistered businesses, startups refer to new businesses that want to grow beyond individual founders.

Eligibility under startup

Under the Startup India program, the DPIIT (Department for Promotion of Industry and Internal Trade) enables eligible companies to access a number of tax benefits, easy to implement, fast track of IPR and many more. A company must meet the following criteria to qualify for DPIIT startup approval and be eligible for all the incentives under various Startup India Advantage schemes.

An entity shall be measured as a Start-up when

- **Type of company:** Whether registered as a private limited company or registered as a partnership or limited partnership in India.
- **Age of the company:** every enterprise up to ten years from the date of establishment/registration.
- **Annual turnover:** If the turnover has not exceeded INR 100 crores for any of the financial years since registration/registration.
- **Innovative and scalable:** If it is beneficial to invent, enhance or refine products, processes, or services, if there exists a business model that can be expanded and the level of anticipated performance or value generation.
- **Parent entity:** Do not create an entity by dividing or restructuring an existing business.

Case-Cost Effective Shade Net House

Profile

Shri Jagnnath Gangaram Tayade, a dedicated farmer from Ladsavangi, Taluka & District Aurangabad, Maharashtra, has been engaged in agriculture for the past 27 years. At 50 years old and having completed his 10th standard education, he cultivates 7.2 hectares of land, growing crops such as cotton, maize, and sweet orange, along with seed production of capsicum, tomato, and watermelon. In addition to crop farming, he also raises livestock, including cows and buffaloes, contributing to a diversified and sustainable farming practice.

Description of innovation

Shri Jagnnath Tayade constructed a shade net house using locally available iron materials. The structure stands 6 meters tall at the center, with a length of 36 meters and a width of 24 meters. Wire ropes are essential for its stability. The design minimizes obstacles within the shade net, allowing for smooth intercultural operations and easy movement of labor, making it more efficient compared to the recommended model.

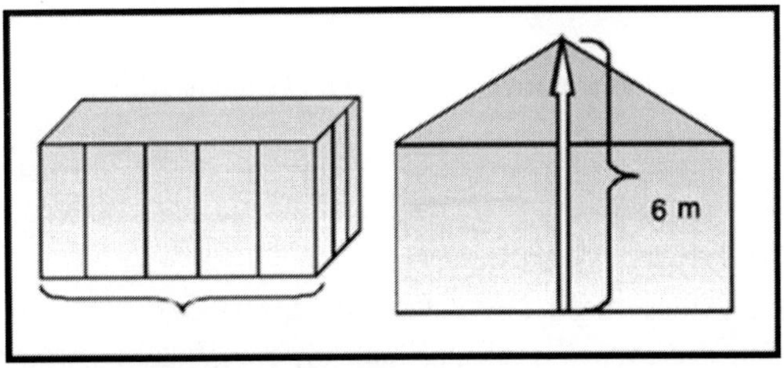

Practical utility of innovation

This shade net house is well-suited for cultivating tall crops such as capsicum, tomato, and chili. Designed to withstand wind speeds of up to 80 km/hr, it is also cost-effective, requiring only ₹60,000 for construction—30% less than the recommended technology. Its dome-shaped structure efficiently drains rainwater, preventing waterlogging. Currently, 15 farmers have adopted this structure on 0.10 hectares each, successfully utilizing it for seed production of capsicum and tomato.

Source: AME Foundation (https://www.amefound.org/)

Benefits under the Startup India

Startups approved by DPIIT can enjoy the following benefits under the Startup India Initiative:

- Self-Certification
- Startup Patent Application and IPR
- Tax Exemption under 80IAC
- Section 56 Exemption
- Easy winding up of Company
- Easier Public Procurement Norms

Let's discuss each benefit in detail

1. **Self-Certification:** To reduce the regulatory burden on startups, so they can focus on their core business and keep costs down.

   Benefits

   - Startups are allowed to confirm their compliance with 6 labor laws and 3 environmental laws through a simple online process.
   - In terms of labor laws, no review is done for 5 years. Alarms can only be investigated if they receive a proper and verified violation complaint, which is in writing and authorized by a higher level of investigating officer.
   - According to environmental regulations, startups that fall under the 'clean zone' (as defined by the Central Pollution Control Board (CPCB)) can demonstrate compliance, and inspections are not carried out. only set up and conducted in these cases.

Labour Laws

- Construction Workers and Other Construction Workers (Laws and Conditions of Employment) Act 1996.
- Lov om interstatlige migrantarbeidere (Laws on Employment and Conditions of Service), 1979.
- The Income Tax Act, 1972.
- The Contractual Activities (Regulatory and Repeal) Act, 1970.
- Wages and Miscellaneous Workers' Compensation Act, 1952.
- The Employees' Insurance Act, 1948.

Environment Laws: Three environmental laws include the Water (Prevention and Control of Pollution) Act 1974, the Second (Prevention and Control of Pollution) (Amendment) Act 2003 and the Third (Prevention and Pollution Control), 1981.

Eligibility: DPIIT recognized startups within 5 years of company registration. Apply for DPIIT accreditation, click "Get Accredited" below.

Registration Process:

- To go to the Shram Sweida portal of the Ministry of Labor and Employment, click on the Incentive Program tab.
- Register at Shram Suvidha portal and login.
- After successful installation, click the link "Are some of your installations still triggering?" click on it.
- Then Follow the instructions.

2. **Startup Patent Application and Intellectual Property Right (IPR)**

Objective of the scheme

- Innovation is the bread and butter of beginners. Patenting is a way to protect new innovative ideas that will make your business competitive, by protecting your product and process it will increase the value and cost of your business.
- However, doing defense in the past has been a costly and time-consuming process that many startups cannot afford.
- The goal is to reduce the cost and time for startups to obtain a patent, while protecting their innovation and encouraging them to innovate.

Benefits

- **Fast-tracking of Startup Patent Applications:** Patent applications from startups need to be quickly screened for scrutiny so that their value can be determined as soon as possible.
- **Panel of organizers to assist in filing of IP applications:** For effective use of the program, a board of "managers" should be established by the Directorate of Patents, Designs and Trade Marks (CGPDTM), which will also manage their work and activities. The coordinators are tasked with

offering general guidance on different types of intellectual property, along with details about the safeguarding and advancement of intellectual property in other nations.

- **Government to bear facilitation cost:** Under the scheme, the government will bear all the costs of the promoters for the number of patents, trademarks or designs registered by the startup, and the startups will bear the legal costs.
- **Rebate on filing of application:** Startups are given an 80% discount on patent applications compared to other companies.

Eligibility: Initiation should be indicated by DPIIT.

Registration Process: For the most up-to-date details regarding the procedures and paperwork necessary for applying for a patent or trademark, you ought to reach out to a suitable assistant - based on the office you desire and the governing body. Follow the link www.startupindia.gov.in/content/sih/en/startup-scheme.html for trademark grantees and patent grantees.

Grievance Redressal: For any queries write to dipp-startups@nic.in.

3. **Tax Exemption under 80IAC**

Benefits: Start-up companies liable to pay income tax for 3 financial years from the first ten years after establishment can be exempted.

Eligibility

- The facility must have an approved DPIIT setup.
- Only private limited companies and limited partnerships are eligible for tax exemption under section 80IAC.
- Must be established after 1 April 2016.

Registration Process and Documents

Registration Process

- Register on the Startup India portal and click on the registration tab.
- Send registration. Apply for the DPIIT (Department for promotion of Industry and Internal Trade) award.
- Get IAC Section 80 Exemption Request Form.
- Fill all the details in the submitted documents and submit the application form.

Registration Documents

- (MOA) Memorandum of Association for Pvt. Ltd. / LLP-tuhinga.
- Board decision (if any).
- Annual accounts of the startup for the previous three financial years.
- Income tax return (ITR) for the previous three financial years.

Process after Applying

- Go to the dashboard on the Startup India website portal to know the status of your application. When you log in, it can be marked on the top right of the screen.
- For follow up, questions and concerns, write to dipp-startup@nic.in.

4. **Section 56 Exemption**

Benefits

- Exemption under section 56(2) (VIIB) of the Income Tax Act.
- Investments in eligible start-ups by listed organizations with net worth exceeding INR 100 crores or turnover exceeding INR 250 crores are exempted under Section 56(2) VIIB of the Income Tax Act.
- Investments in start-ups can be made by qualified investors, non-residents, AIF (Category I) and listed organizations with a capital of more than INR 100 crores or more than INR 250 crores is exempted under Section 56(2) (VIIB) Income Tax Act.
- Consideration of shares acquired by authorized organizations up to INR 25 crores is exempted.

Eligibility

- Must be a private limited company.
- Requires active DPIIT control. To get DPIIT qualification click on "Get Approval" via URL https://www.startupindia.gov.in/.
- Does not invest in certain asset classes.
- The start-up should not invest in immovable property, vehicles above Rs.10 lakh, loans and advances, capital contributions to other entities, except in normal business activities.

Registration Process

- Register your startup profile on India startup portal.
- Receive DPIIT certification.
- Fill Section 56 Exemption Application Form www.startupindia.gov.in/content/sih/en/Form-56.html.
- You should receive an email from the CBDT within 72 hours of submitting your return.

5. **Easy winding up of Company**

Objective

- Make it easier to simplify or eliminate the activities of entrepreneurs, with the goal of enabling entrepreneurs to more quickly allocate capital and assets in productive ways.
- Encourage entrepreneurs to experiment with new and innovative concepts, without having to deal with cumbersome and long exit procedures in areas where their capital will be lost if the business fails.

Benefits

- According to the Insolvency and Bankruptcy Act, 2016, start-ups with a simple credit structure, those with real income can be discharged from the * criteria within ninety days of filing for bankruptcy.
- A free agent is to be designated to set up the company and oversee its operations going forward (the promoters and directors are not responsible for managing the company) which includes liquidating its assets and settling his debts within six months.
- Following the selection of a liquidator, the liquidator will take charge of the prompt liquidation of the business, asset sales, and creditor payment in accordance with the liquidation procedure initiated under the IBC. This approach can honor the notion of limited liability.

Eligibility and Process: Visit: http://www.ibbi.gov.in/

6. **Easier Public Procurement Norms**

Objective

- Public procurement refers to the way in which government and public agencies purchase goods and services from the private sector.
- Public procurement refers to the way in which government and public agencies purchase goods and services from the private sector.
- Government organizations possess significant funds to utilize, making it an excellent market to begin with.
- The goal is to make it easier for companies to participate in public procurement and to reach new markets for their products.

Benefits

- Ability to list your goods on Government e-Marketplace: Government e-Marketplace (GeM) is the largest online shopping platform and marketplace for government agencies to purchase products and services. . Startups recognized by DPIIT can register as traders in GeM and directly supply their goods and services to government agencies. This is a great opportunity for businesses to work with authorities on pilot orders.
- Exemption from previous experience/turnover: To increase startups, the government exempts manufacturing companies from the "previous experience/turnover" criteria without compromising quality standards and technical parameters. has been announced. Activists must also demonstrate that they have the necessary capabilities to complete the project on schedule and within the construction period.
- EMD Waiver: DPIIT does not require startups to provide an EMD (EMD) and security application when bidding on government contracts.

Eligibility: The start-up must be notified by the Department of Business Promotion and Internal Trade. For more information visit https://www.startupindia.gov.in/content/sih/en/Startupgov/startup-recognition-page.html.

Grievance Redressal : For queries write to dipp-startups@nic.in.

Scenario of Startups in India

According to the Economic Survey 2021-2022, more than 61,400 startups in the Asian country have been recognized by the Department for promotion of Industry and Internal Trade (DPIIT), and at least 14,000 have been identified across the industry by 2022. This survey is an annual economic report that examines the performance of various sectors and predicts future trends. This includes forecasts of GDP growth In the study, 555 regions of the Asian continent have less than one new start-up, indicating that start-ups have grown significantly in the countries of Asia over the past six years, mostly in the IT/knowledge sectors. (www.startupindia.gov.in).

"Startups in India have grown exponentially in the last six years. The number of newly recognized startups has increased from just 733 in 2016-17 to over 14,000 in the 2021-22," the study said. According to this report, India has become the third largest startup ecosystem in the world after the US and China. According to the survey, a record 44 Indian startups reached unicorn status by 2021, taking the total number of unicorn startups in India to 83, most of which are in the service sector. The survey also revealed a series of IPOs from new companies.

"During April-November 2021, Rs 89,066 crore was raised in seventy-five IPOs, more than any year in the last decade," it said. The study also parallels the expansion of tech startups in India. Based on this research, the quantity of startups in this industry has risen from eleven in 2019 to forty-seven in 2021. In recent years, Delhi has replaced Bengaluru as the capital of India. There are over 5,000 popular startups (Economic Outlook 2021-22) Opp starts scenario in Gujarat: (https://startup.gujarat.gov.in/home).

Gujarat seems to be known for its entrepreneurial identity, which makes this mixed state one of the most sought-after start-up locations. In Gujarat, there is a joint project called Startup Gujarat. Startup Gujarat can be a place to collect, evaluate and promote innovations that can help you start your business.

- Registration of more than 8450 startups and support to more than 180 centers or incubators across the state with support of 200 INR core.
- Creation of more than 100 defense data centers in various academic institutions.
- More than 900 registered patents with government support.
- Creation of more than 30 scientific community buildings at the local and regional level.
- Support incubators and centers to organize various programs, including boot camps, hackathons, accelerator programs, etc.
- "Prime Minister's Award for Excellence in Public Administration 2017", for implementation of Startup India program.
- "Best expert" in the national leader for startups in 2018 and 2019.

Government Programs under the Startup India Initiative

Government Programs under the Start of India Under this flagship initiative of the Government of India. This Indian startup currently has different variants available

(Table 9.1). As mentioned earlier, which of the front lines of the Indian start-up, along with the many benefits planned for this sector, is provided by the project.

Table 9.1: Government initiatives under the Startup India initiative

Sr. No	Name of the Schemes
1	Venture Capital Assistance Scheme
2	Support For International Patent Protection in Electronics & Information Technology
3	Stand-up India
4	Single Point Registration Scheme
5	Extra Mural Research Funding
6	High-Risk High Reward Research
7	IREDA NCEF Refinance Scheme
8	Dairy Entrepreneurship Development Scheme
9	Revamped Scheme of Funding For Regeneration of Industries
10	Assistance To Professional Bodies & Seminars/Symposia
11	Scheme To Support IPR Awareness Workshops/Seminars In E&IT Sector
12	Biotechnology Ignition Grant
13	Small Business Innovation Research Initiative
14	Biotechnology Industry Partnership Programme
15	Pradhan Mantri Mudra Yojana
16	4E(End To End Energy Efficiency)
17	Sustainable finance scheme
18	Promoting innovations in individuals, start-ups and MSMEs
19	NewGen Innovation And Entrepreneurship Development Centre
20	International cooperation scheme
21	Enhancement of competitiveness in the Indian capital goods sector
22	Bridge loan against generation based incentive claims
23	ASPIRE
24	Ayurvedic biology Programme
25	Technology development Programme
26	Software technology Parks team
27	Self-employment lending schemes-Credit Line 2- Term Loan Scheme
28	Virasat a credit scheme for Crafts persons
29	Self-employment lending schemes- Credit Line 1- Term loan scheme
30	Self-employment lending schemes – Credit Line 1-micro financing scheme
31	Self-employment of lending scheme - Credit Line 1- Mahila Samriddhi Yojana
32	Self-employment lending scheme -credit line 2- micro financing scheme
33	Coir Udyami Yojana
34	Informediary services

35	Raw material assistance scheme
36	Coir Vikas Yojana
37	Credit facilitation through bank
38	NSIC Infrastructure scheme-IT Incubator
39	Joint advertising support by Ministry of tourism
40	Brochure support by Ministry of tourism
41	Marketing assistance scheme
42	Procurement and marketing support scheme
43	Time share resort
44	Travel trade
45	National awards (Individual MSEs)
46	National manufacturing competitiveness programme
47	Composite loan
48	Performance and credit rating scheme
49	Marketing development assistance for promotion of domestic tourism
50	Marketing development assistance for promotion of domestic tourism for medical/ wellness tourism
51	Assistance under scheme for setting up tea boutiques in India for domestic Promotion
52	Swarojgar credit card
53	Material testing labs
54	Self employment and talent utilisation
55	Micro exporters policy
56	Export development and promotion of spices
57	Reimbursement of certification fees
58	Marketing support and assistance to MSMEs (Bar code)
59	Lean manufacturing competitiveness for MSMEs
60	Entrepreneurial and managerial development
61	Amended technology up gradation fund scheme
62	Venture capital by SIDBI Venture capital Ltd.
63	Tex venture fund
64	Samriddhi fund.
65	Export promotion capital goods scheme
66	Central sector scheme for promotion of international cooperation in AYUSH
67	Credit guarantee fund trust for micro and small enterprises
69	The design clinic scheme
70	Science and Technology (S&T) for Coir Institutions
71	Research and development in processed food
72	Creation / Expansion of Food Processing & Preservation Capacities

73	The Women Entrepreneurship Platform
74	National Urban Livelihoods Mission
75	Electronics Hardware Technology Park Scheme
76	Skill development initiative
77	Udaan training programme for unemployed youth of J&K
78	Duty exemption and remission scheme
79	Support to training and employment programme for women
80	Comprehensive Handloom Cluster Development Scheme
81	Apprenticeship training
82	Trade Related Entrepreneurship Assistance and Development scheme for women
83	Cent Kalyani scheme
84	Marketing intelligence services
85	Prime Minister 's employment generation programme
86	National Skill Certification & Monetary Reward
87	Pradhan Mantri Kaushal Vikas Yojana
88	Craftsmanship training
89	Skill Development in 34 Districts Affected by Left Wing Extremism
90	Upgradation of 1396 ITIs through PPP
91	Agro processing clusters scheme
92	Apparel Park
93	Research and Development of textiles
94	Rental for warehousing
95	Promotion of packaged tea of Indian origin
96	Participation in International Fairs and Exhibitions with Tea Board
97	Bridge Loan against MNRE Capital Subsidy payable to Accredited Channel Partners and State Nodal Agencies (SNA) for claims against installation of Solar Water Heating Systems
98	Atal Mission for Rejuvenation and Urban Transformation.
99	Drugs and pharmaceutical research
100	Mega clusters
101	Development/Up gradation of Watermills and setting up Micro Hydel Projects (upto 100KW capacity)
102	Research, Design, Development, Demonstration (RDD&D) and Manufacture of New and Renewable Energy
103	Wind resource Assessment in uncovered /New areas Under NCEF Scheme
104	Capital investment subsidy scheme under NB
105	Capacity building and technical assistance
106	Corporate finance (North-Eastern Development Finance Corporation Ltd.)
107	Equipment finance (North-Eastern Development Finance Corporation Ltd.)
108	Initiative for Development of Entrepreneurs in Agriculture

109	Micro Finance Scheme
110	Equity funds scheme
111	NEDFi Opportunity Scheme for Small Enterprises
112	North-east entrepreneurs development
113	North-east handloom handicrafts
114	Rupee term loan
115	Working capital term loan
116	WCTL for Contract Finance
117	Women enterprise development
118	North-Eastern region urban development programme
119	North-east rural livelihoods project
120	Advertising and Publicity by the Ministry of Development of North-Eastern Region
121	Central sector scheme for MSMEs
122	Technology development fund
123	Digital communication innovation Square

Basics Of A Business Plan

The foundation of a business plan revolves around an idea and the vision of an entrepreneur. It is the entrepreneur's dream and vision that has to be pursued in the light of various ideas that can be said as having market potential at a given point of time that gives shape to the business plan.

A business plan includes the following major topic headings:

- Introductory page
- Executive summary
- Vision/mission statement
- Business environment and competitive analysis
- Market research and analysis
- Marketing plan
- Financial plan and funding needs
- Technology, manufacturing and operational plan
- Intellectual property
- Legal considerations and insurance requirements
- Management team
- Risks, problems and assumptions
- Launching a business and measuring plan progress
- Growth and exit strategy
- Appendices

A complete business plan with a strong foundation needs to be prepared by way of clarity on the aspects just mentioned, which need to be aligned with the vision and mission of the entrepreneur for the business idea. Further, it is not enough to prepare a good plan, but it should also have an effective implementation strategy in place. Above all, what matters most is to effectively communicate the plan to different stakeholders, which requires a knack of making elevator pitch, brief presentations and detailed presentation skills.

1. Introduction

The 'Introduction' to a plan needs to be presented in around a page and should briefly cover details of the legal form of the business entity-company, partnership and sole trader. One should precisely explain in one or two sentences what the business does or plans to do and what is the purpose of the plan.

Executive Summary, Vision/Mission, Business Environment and Competitive Analysis, Market Research and Analysis, Financial Plan, Marketing plan, Technology, manufacturing and Operational Plan, Intellectual Property, Legal Considerations, Insurance Requirements, Management Team, Suppliers, Risks, Problems and Assumptions, Proposed Venture Offerings, launching a Business, Growth and Exit strategy, Appendices

2. Executive Summary

Last to be written and first to be presented is the executive summary of a business plan. A precisely well-written executive summary acts as a critical factor to the success of a business plan. This section of the plan should be so catchy and powerful that it attracts the attention and interest of the reader to get compelled to read the full plan. It is important to note that each and every word used in the executive summary should be carefully selected and presented. This section should accurately define the nature of the business venture, including the customer need that the business proposes to fulfil through innovative products or services. It should emphasize on why people would need it, highlighting a brief analysis of the characteristics of potential market. It provides a precise and attractive gist and highlights the key points related to the product/service, funds required, returns expected, market potential, the team backing it and their commitment. It should distinctly highlight to the prospective investor the size and scope of the market opportunity, the venture's business and profitability model and how the resources/ skills/strategic positioning of the company's management team make it uniquely qualified to execute the plan. The executive summary is a powerful tool that attracts the attention of stakeholders in general and investors in particular and should not be presented in more than 2 to 3 pages. It should include the following (Fig.9.1):

- A brief description of the venture
- The objectives of the business
- A brief description of the company's products or services in relation to the identified customer needs
- Target market and competition

- Competitive advantage of the business and rationale for why the business would succeed
- Growth prospects
- Strength of the management team
- Funding requirements, including proposed schedule for fund deployment

Figure 9.1: Key Aspects of Executive Summary

The executive summary should be brief is not more than two to three pages and written in powerful words to generate interest on the part of stakeholders in the venture. It briefly covers the fundamentals of the proposed business: What will one's product be? Who will one's customers be? Who are the owners? What does one think the future holds for one's business and one's industry? What would be the fund requirements at different points of time and how would one use them? What would be the returns on equity? It should be made interesting, fascinating, professional, concise, complete and highly communicative in simple language.

3. Vision/Mission Statement

Every successful entrepreneur has a dream. It is this dream that the entrepreneur attempts to translate into reality. The basic purpose of existence of a business entity and what it does needs to be clearly defined in simple terms by the entrepreneur. Entrepreneurial leadership, as discussed earlier, requires a clear vision, that is, a potential that an entrepreneur wants to unfold an ability to clearly communicate the vision to all those who matter in order to realize the same, and an ability to inspire and motivate all those people who would happily and willingly contribute their best of resources to enable the realization of the vision.

The vision statement for the venture basically refers to the long-term vision of a company, highlight- ing where it would like to be in another 10, 20, 30 or even 50 years from now. It presents a picture that is similar to a dream for the organization. It charts a strategic path for the future as to 'Where are we going?" and steers the energies of employees towards a greater purpose for which the organization exists. It has to be challenging. It should provide a direction for the venture that needs to be focused, flexible and feasible to achieve.

A mission statement should clearly define-what-the-business wants to do mainly for customers, employees and owners. A 'mantra' is a single phrase that defines a business. Guy Kawasaki, author of Art of the Start, recommends a mantra instead of a mission. A mission statement highlights, "What does the business do?"; "What business are you in?". It basically provides a purpose for its existence by way of what it does. It gives the core purpose for the existence of a business and its identity by focusing on current product and service offerings, customer needs being served, and technological and business capabilities. An effective mission statement can be developed by getting inputs from all members of the organization. It should be precise within two to three sentences. Above all, an entrepreneur should necessarily believe in a mission statement; if one does not, then other stakeholders including customers would soon realize it.

4. Business Environment and Competitive Analysis

Business environmental analysis is the first step towards getting the broad clarity about a business environment into which an entrepreneur would like to plunge. It is also called PEST (political, economic, sociocultural and technological). As such, two other vital components of PEST include legal and ecological ones. This analysis helps a lot in chalking out strategies to take advantage of opportunities and to make contingency plans for minimizing the adverse impact of threats.

PEST analysis aids in recognizing important political elements or legislative modifications that might influence the business. In case a venture is spread over multiple countries, then one needs to examine the implica- hons of the political system and policies in each country. The political factors include aspects such as licensing, flow of capital-related policies, laws on maternity rights, data protection and environmental policy.

Although tax rates are typically established by politicians, tax decisions typically incorporate economic factors like growth rates, poverty levels, and the industrial landscape. Additional economic elements that must be taken into account include exchange rates, inflation rates, income growth, debt and savings, as well as consumer and business confidence. For specific industrial activities, one might need to emphasize particular industry metrics that become significant, such as the availability of a skilled labor pool, labor regulations, or the costs of raw materials that would have varied business consequences.

Certain commercial enterprises have significant consequences stemming from technological advancements and, therefore, if these businesses do not adapt to new technologies, it could lead to their downfall. Technological developments also influence political and economic factors, and play a part in how people view their world. For example, Internet technology has made a major impact on the way consumers make their decisions to market and businesses undertake marketing research and purchase raw materials. As such, even a small business can now serve a global market and can even undertake businesses that may not require an office space or a distribution network. Environmental factors to be examined, particularly for certain businesses that are environmentally more sensitive, include the impact of climate change.

The changes in the sociocultural profile and factors create business opportunities, on the one hand, and make some of the existing businesses redundant, on the other. Social factors influence and manipu late people's choices and include societal beliefs, values and attitudes. One needs to examine demo- graphic changes and also consumer views on one's product and industry, lifestyle changes and attitudes to health, wealth, age, gender, work and leisure.

A good environmental analysis helps in a better understanding about the industry in which the busi- ness would fall. Industry analysis, examines and evaluates the playing field in which the company is competing, and attempts to get a clearer picture about the size of the market. It also helps in identifying genuine market needs to assess the market potential by clearly identifying relevant market size. Some of the fundamental questions that industry analysis helps in answering are given as follows:

- What are the sizes of the target relevant market segments?
- What have been the past and emerging trends for the industry as a whole?
- Who are the competitors with whom one's products/services have to compete?
- Is the product/service under consideration just a fad or will it provide a long-term sustainable market?
- Are we talking about a seasonal market? If so, what is its period and intensity?
- Is the industry growing at an accelerated, constant or decreasing rate? Or, does it just grow with the growth of the population?
- What are the main elements that would affect the expansion of the market and of market shares?
- Who are the potential customers for the business product or service?

Industry analysis additionally aids in recognizing profit margins that differ significantly from one industry to another. For instance, margins might reach up to 24 percent in mining and crude oil extraction, approximately 15 to 17 percent in pharmaceuticals, 6 to 7 percent in food and consumer goods, and as low as 1 to 1. 5 percent in automotive retail services.

Competitive Analysis This section of the business plan outlines the competitive environment surrounding one's business. It pinpoints both direct and indirect competitors and evaluates their advantages and disadvantages. This assists in recognizing the primary competitive strengths of one's business that should be prioritized. If one notices that current market participants are facing significant challenges, then it's essential to determine the causes, in order to make a well-informed decision on whether entering the business would be beneficial or whether to avoid repeating past errors. If existing market players are performing exceptionally well, then it may be necessary to provide a justification for introducing another competitor into the market. Therefore, competitive analysis involves the examination of competitors' business models, product specifications, customer demographics, performance, and marketing strategies in the marketplace. This aids in formulating an effective marketing plan that includes product features, packaging, pricing, distribution channels, quality, and service to achieve a competitive edge. One must evaluate all close competitors based on factors such as the competitor's objectives, their products/services, and their competitive strengths through their operational capabilities and strategies in the market.

The specific aspects that must be covered in this section of analysis are described as follows:

- Identify one's closest competitors. What are their key strengths and weaknesses? Where are they located? What are their revenues? How long have they been in business?
- Define their target market.
- How is their business doing? Is it growing? Is it scaling back?
- What is their share off market?
- How is one going to be different when compared with one's close competitors?
- What are the areas of improvement that can provide an extra edge to one in the market?
- In what ways is one's business superior to the competition?
- How are their operations similar to one's business and how do they differ?
- Are there certain areas of the business where the competition surpasses one? If so, what are those areas and how does one plan on compensating?

Competitive analysis places customer needs and desires at the core of the evaluation. An entrepreneur, upon assessing the offerings provided by rivals, should regard customer priorities as their business priorities. The entrepreneur's entire focus shifts to creating innovative products and services that are essential for achieving business success in any sector. It is the responsibility of an entrepreneur to foresee upcoming customer needs and desires, ensuring they adapt to evolving customer requirements by suitably enhancing and refining product offerings.

Abrams (1993) cited several common barriers to entry for new competition

- **Patents** - These offer a degree of protection for innovative products or processes.
- **High Start-up Costs** - Often, this obstacle poses the greatest challenge for small enterprises.
- **Knowledge** - A deficiency in technical, manufacturing, marketing, or engineering skills can present substantial hurdles to successful market entry.
- **Market Saturation** - It is a fundamental truth that it is more challenging to find a niche in an overcrowded market compared to establishing a foothold in a market characterized by comparatively low competition.

'In reality, only a few entry barriers endure for an extended period, especially in emerging industries, Abrams determined. "Even patents offer far less protection than is typically believed. Therefore, you should realistically estimate the timeframe in which new rivals will overcome these obstacles.'

On the quality front, one may undertake a detailed analysis on various parameters of quality for different competitors to identify others' areas of weaknesses, so as to create a distinct niche. The characteristics could be performance-related, feature-related, durability, adherence to pre-established standards, ease and speed of service, packaging and aesthetic-related. Service quality parameters could be timeliness, courtesy, convenience, accuracy and consistency. Another way in which one can precisely work out competitiveness is by rating on a scale of 0 to 10 the various attributes of one's product vis-à-vis competitors. If needed, various attributes can be given different weightage, depending on their importance in the mind of customers. This would help in getting overall attractiveness or unattractive- ness of one's product in comparison to competitors and will also help identify key attributes that would require greater focus to excel in the market (Fig.9.2) .

	Your Company	**Competitor # 1**	**Competitor # 2**
Market Share	x	x	x
Cost	x	x	x
Key Product Attributes	x	x	
Brand Image and Reputation	x		x
Leverage With Suppliers			
Technology	x		

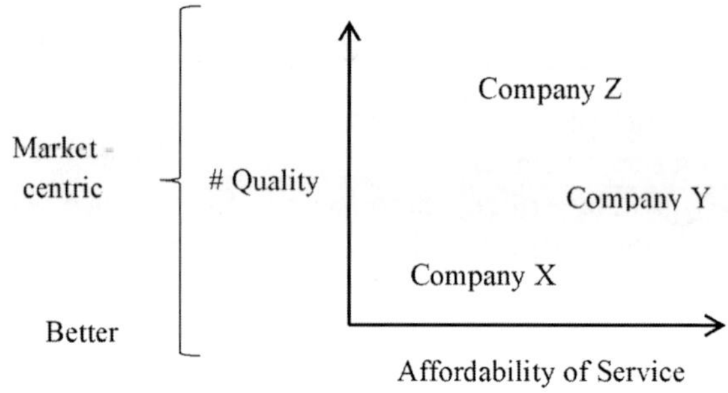

Figure 9.2: Examples of Competition Graphics

This approach is called "company-centric'. One can work out competitiveness from the point of view of the market-centric approach on different parameters to identify one's company's niche in comparison to the competitors. It is generally advisable to be market-centric to identify key strategies that would help one's venture succeed.

In spite of having scientific tools and techniques undertake competitive analysis, it is important to remember that understanding and predicting competitors' actions is a complex phenomenon. Competitors, as such, do not always behave in a rational manner and would like to always retain suspense and uncertainties about their strategies and actions. This complicates the process of competitive analysis and makes it difficult to forecast rivals' actions. This requires building alternate strategies to counter competitors' moves, thereby resulting in scenario building,

Analysis of one's competitors should be a continuous process, so as to avoid any sudden shocks. Having a clear understanding about competitors helps one in becoming more motivated to succeed, efficient and effective in the marketplace.

High-technology-driven ventures that change the whole profile of an industry may lead to driving first-mover advantage and, in turn, the least of competition to begin with. However, even these ventures have a risk of competition emerging over a time period or with the allied substitute products/services. Mostly research and development efforts emerging out of universities and research centres may result in breakthrough technologies, thereby giving rise to competitive advantages.

5. Market Research and Analysis

The main purpose of market research and analysis is to assess the various customer segments that the venture would serve. This section clearly includes the customer needs that a business venture proposes to satisfy. It should also justify the value proposition vis-à-vis the price that a customer will willingly pay for the product/ service. The main purpose of conducting market research and analysis is to prion. tize target markets and get a clearer picture about customers and market segments. One needs to dwell on issues such as demographic and psychographic characteristics of one's target customers. What can be the best mode to approach and reach them? What are their main concerns? What approach would they like to reach them the most?

It is important to consider the potential market for one's product/service and not the actual market captured or served at present. A target market provides one with a picture about the scope of the market that can be captured and that one needs to be concerned with as an entrepreneur. For example, the market of a local club or a taxi operator includes not only the customers who are availing the service restaurant and the people who are regularly availing these services but also all those who could avail these services. Similarly, a college having 10 degree programmes with a capacity of 500 students' intake each year and that is looking for improving the quality of student intake may have to look at all those students who are eligible for securing admission in the college, but who do not apply at present, as target market. Similarly, the market for downloadable e-books over the Internet includes all those who have Internet connectivity. The market for laptops includes homes, schools, businesses and government organizations.

Analysing the market from a chronological perspective is also desirable, thus highlighting a historical market survey giving details about product sales and competitors, present market status, and trends and forecasts for the future from a growth perspective.

Market research and analysis should fundamentally enable one to answer questions such as: What is break-even sales and when will it be achieved? Is the market potential high and good for sustain- able growth? Does one have something rare and distinct to offer, so that one can derive competitive advantage?

As suggested by Malcolm McDonald, the key aspects to be covered under marketing research and analysis are described in the following (Fig. 9.3):

- **Understanding Marketing Environment-**A study of markets, customers, competitors and the overall economic, political, cultural and technical environment covering emerging trends and present situations is analysed in detail to understand market prospects for the product/service in a given identified industry.
- **Develop a Detailed Marketing Mix** - The term 'marketing mix was coined and first used by Neil Borden in 1953 in his American Marketing Association presidential address. McCarthy, a prominent marketer, proposed a four-P classification in 1960, which has seen wide-ranging applications in the marketing literature." The four Ps concept of marketing mix are as follows:
- **Product** - A tangible object or an intangible service with its key ingredients and features that are produced to cater to customer needs. Intangibles include services like hotels and the tourism sector, educators, medical professionals, and attorneys, while tangibles encompass items such as mobile phones, televisions, automobiles, furniture, loans, and credits. It is crucial to keep in mind that each product goes through a life cycle featuring stages of introduction, growth, maturity, and decline. To stay competitive, an entrepreneur must distinguish their products and continuously innovate new offerings to launch them before the decline phase of the current products starts.

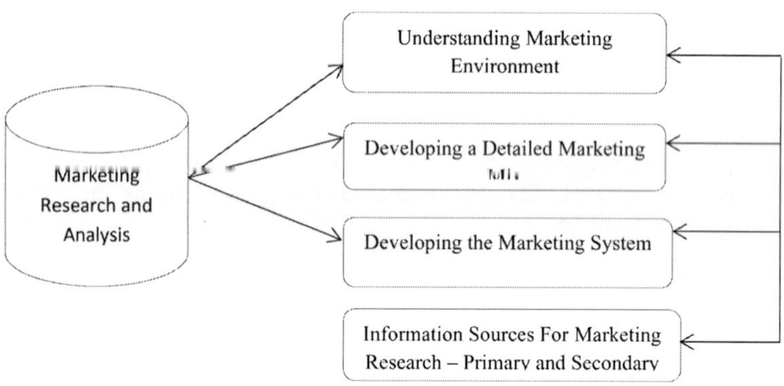

Figure 9.3: Aspects of Marketing Research and Analysis

- **Price** - Entrepreneurial endeavors necessitate an important choice concerning price determination, which relies on several elements including market share, rivalry, production costs, product branding and identity, and most importantly, the customer's perceived value proposition.
- **Place** - This concentrates on the place where a product or service is offered for purchase by the customer. It is referred to as the 'distribution channel' employed by the company to facilitate the buying process. It may involve both a physical and a digital process to provide the product/service to the customer.
- **Promotion -** This process pertains to the communication methods, instruments, and strategies employed by the company in the market to make a product recognizable to the intended customers. It includes advertising, public relations,

personal selling and sales promotion. Any organization, before introducing its products or services into the market, conducts a market survey. The sequence of all Ps, as just mentioned, is very important in every stage of product life cycle: introduction, growth, maturity and decline.

Subsequently, three more Ps have been added to the marketing mix, namely, people, process and packaging.

- **People** - It is most important to understand well all those who are involved in producing and consuming such as customers, employees, management and, at times, even the community, in terms of their relevance for relation building. This aspect of a marketing mix also facilitates the proper identification of market segments, that is, a class of people from demographic, economic, social and psycho- logical point of view for whom the product or service is to be made available.
- **Process** - The procedure, mechanism and flow of activities through which a customer is reached by the manufacturer.
- **Packaging** - This focuses on how a product is protected and presented and the manner in which a service or product is delivered to the customer - the tangible ones help to communicate, and the intangible is the knowledge of the people around us.
- **Developing the Marketing System** - A study of the marketing organizations, marketing research systems and the current marketing objectives and strategies. The last of these is too frequently ignored. The marketing system itself needs to be regularly questioned, because the validity of the whole marketing plan is reliant on the accuracy of the input from this system, and garbage in, garbage out' applies with a vengeance.
- **Information Sources for Market Research** - There are two main categories through which information for market research can be gathered, specifically, primary sources and secondary sources. Primary data is obtained from original sources through first-hand accounts that act as foundational elements for market research objectives. These are called 'primary sources', as they are the first evidence of something happening and are collected directly from the source when the event is happening or immediately after the event has happened. Some of the examples of primary data sources are diaries, letters, interviews, questionnaires, video/films, newspapers and outcome of experiments. Secondary data sources are the information collected from secondary sources, as they are created after primary sources, and they often use or talk about primary sources. For instance, information gathered from the chamber of commerce, governmental reports, journals, research documents, and the web. Secondary sources frequently exist in multiple copies that can be located in libraries, educational institutions, or residences, such as information compiled from textbooks, biographies, published materials, and governmental reports. The general differentiation between primary sources and secondary sources is provided next.

6. Marketing Plan

Investors seek management teams that have a distinct comprehension and awareness of their target market. While launching a new venture, a marketing plan based on marketing research should be developed. In case of a venture having existing customers, the entrepreneur should provide an analysis of who one's customers are, their purchasing habits and their buying cycle. A marketing plan based on marketing research and analysis describes sales projections and the mechanism to achieve them. It attempts to develop concrete estimates for sales and a strategy that will enable the entrepreneur to avail the opportunity in the light of specific competitive advantages. The whole purpose is to convert an estimated physical quantum of sales and the corresponding sales revenue that is expected to be achieved. It is essential to recognize that a successful marketing plan might not need a substantial amount of funds, but it undoubtedly demands creativity and resourcefulness to develop effective strategies that would produce desired outcomes. A network of connections helps a lot in the entrepreneurial journey as a whole and, in particular, for start-up ventures. A sound and robust marketing plan would reflect a better understanding of customer needs, and it recognizes that business success revolves around satisfying customers and their needs. It helps in developing a strategy for success, purely from customers' perspectives. A marketing plan has to clearly describe what is to be done to achieve sales targets, the strategy to do this, when different activities under the marketing plan will be required to be undertaken and who will do what, including review and necessary corrective actions. Some of the key aspects to be covered in the marketing plan are given as follows:

- Description of one's target market.
- Focus of one's business in the given target market from the niche one proposes to create for capitalizing on a market.
- Realistic assessment of the niche market with a basis for estimation.
- Documents to support why one believes there is a need for one's product or offering in this market.
- Percentage of market share that one proposes to capture and its rationale.
- Estimated growth potential of the market.
- Expected increase or decrease in one's market share with expected growth in the overall market.
- The steps that are required to be undertaken to retain or increase one's market share in the overall growth of the market.
- Mechanism to set prices for the products in the growing competitive market.

A marketing plan having other than the details just mentioned should also cover the entry barriers in the industry such as high capital costs, high marketing costs, brand recognition, human resource competence, unique technology and patents, tariff barriers and quotas that would give one's venture a unique advantage against competitors.

7. Financial Plan and Funding Needs

A financial plan assesses the financial implications of various strategies proposed in other sections to quantify the flow of profit and cash vis-à-vis investments required. All other aspects get converted into financial projections in terms of both funding requirements and income-generating capacity of the venture to generate adequate return on investments. A business plan should be able to convert various numbers into the financial plan and funding needs of the venture for one's firm. One of the key purposes of writing a business plan is for investors to raise funds. Therefore, it necessarily requires an estimation of fund requirements and sources from where funds are proposed to be raised. The business plan effectively communicates the proposed roadmap laid out by one to potential investors and other funding agencies about one's business.

The extensive funding sources might include a mix of the promoter's equity, equity investments from additional owners, retained earnings reinvested into the business, subsidies and grants obtained, and various forms of financing including both short-term and long-term loans. Thus, the three broad categories of funding could be debt, equity, and grants and subsidies (Fig. 9.4). As a business grows, it may necessitate the need for additional funds. The venture may look for private equity funding, venture capital funding, IPO, private lenders, banks and financial institutions.

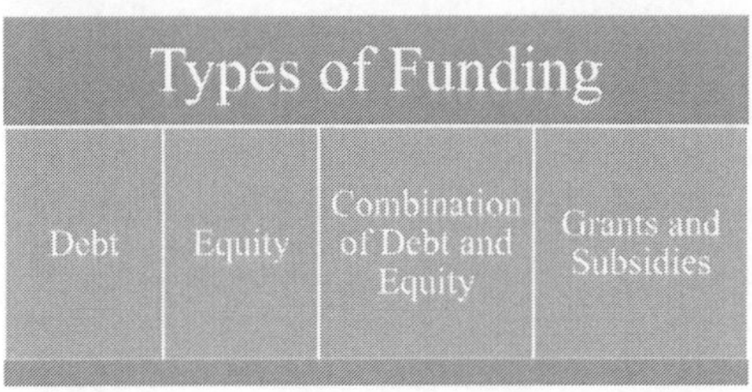

Figure 9.4: Categories of Funding for Ventures

Any investor would be keen to know about and understand the various components of one's business plan, namely, the marketing strategies proposed to be used, the operational plans, human resource strategies and management team. Above all, investors would seek to grasp the final figures, which would be clear from the business financial plan, as this would serve as the primary gauge of the well-being of one's enterprise. This requires making realistic assumptions to work out the projected profit and loss account, balance sheet, fund flow and cash flow. An entrepreneur should forecast sales and expenses expected at least within the coming three years of the projected profit and loss account with monthly projections for the first year. Sales forecast, cost of goods sold and other fixed expenses such as administrative, depreciation and interest are required to arrive at profit before taxes. Net profit could be arrived at by projecting taxes payable. The second key information needed is cash flow projections for three years, with the first-year projections being monthly. Since a business has to pay its bills on time, which is essential for its survival and growth, one should precisely estimate the requirements of funds at different points of time. It is essential to keep in mind that a business's offerings should possess an influence over consumers to receive payment either beforehand or at the point of delivery, ensuring that the forecast of sales income does not become a significant issue. Usually, when a venture does not have an extraordinary edge in the competitive market scenario, the realization of sales may require providing some credit terms that result in irregular receipt patterns from customers, necessitating borrowing or induction of funds by the entrepreneur to manage payment of fixed expenses and salaries on time. Having prepared a projected profit and loss account and cash flow, one needs to prepare a balance sheet of the company that highlights the financial position of the company at a given point of time. A balance sheet basically gives the assets position that is, what is owned by the company, and liabilities, what is owed to others, including investment by the entrepreneur and other equity holders and retained earnings or cumulative losses. Any key assumptions made in the financial plan pertaining to working out the projected balance sheet should be listed separately for the benefit of understanding by different stakeholders and, in particular, for the benefit of investors.

8. Technology, Manufacturing and Operational Plan

A business plan for products or services should necessarily include an operational plan highlighting key aspects such as product or service characteristics; production, planning and control aspects; location, equipment, people, processes and surrounding environment. It is essential to remember the basic distinction between physical goods and non-physical services. Therefore, in the case of services, characteristics like taste, touch, feel, and sound are highly significant. For instance, hotels, legal services, vehicle rentals, and healthcare facilities greatly depend on intangibles for their business success. The method of providing the services essentially sets one business apart from another in a sector of comparable services.

The description of products and services includes features, content, and usefulness of products and services from a customer's perspective. An entrepreneur should describe specific and important features and special aspects about the products and services offered by them and their benefits to the customers.

This section covers the methods used for producing products and services highlighting production techniques and costs, quality control; customer care and service; inventory control and product development aspects. It should also cover locational aspects of the venture, highlighting specific requirements that the venture would require from a locational advantage point of view by giving details of space required, type of building, availability of power and other utilities. The accessibility of a site concerning transportation, suppliers, and customers must be carefully assessed. Should there be particular needs regarding parking and closeness to freeways, airports, railroads, and shipping centers, these should be properly evaluated and addressed when choosing a location. One has to take care of the aspects related to drawing and layout of the proposed facility. It is advisable for start-ups not to invest heavily in building construction and manage the initial phase by taking buildings on rent or lease, unless and until it is highly essential for the venture to own a building. Briefly, this section of the business plan should focus on the following key aspects:

- Is one's product or service already on the market having competitors or is it still in the research and development stage?
- In case one's product or service is still in the development stage, what is the roll-out strategy or time- frame to bring the product to the market?
- What are the unique aspects of one's product or service? What are the key competitive advantages that one's product or service has over its rivals?
- Is it possible for one to price the product or service competitively and maintain a healthy profit margin?
- What methods are proposed to manufacture/produce or provide services for the market?
- What technologies will be required and what is the status of their advancement?
- At what pace are new technologies in the area emerging that result in making existing technologies obsolete?
- What production, planning and control systems does one propose to introduce and what are the problems anticipated?
- What essential elements are linked to the achievement of an individual's production/manufacturing/servicing operations?
- What is the criticality of different factors of production and processes?
- How does one propose to handle these critical factors?

9. Intellectual Property

Start-up ventures having great potential in raising venture capital have a wealth by way of proprietary intellectual property (IP). As such, two key aspects looked into by investors are the management team and the quality of the IP that can give rise to a unique advantage to the venture. In a business plan, together with the description of the key features of one's product, an entrepreneur should specifically highlight the patents acquired in different countries vis-à-vis market potential in those countries in which the patents have been acquired. A business plan ought to concentrate more on the advantages and uses of the IP instead of the confidential elements of the IP. It is crucial to keep in mind that even the most groundbreaking technologies might not capture investors' interest unless they present concrete benefits to the customers.

The business plan must effectively emphasize the method of integrating the IP with the suggested products and services provided. It may be advisable at times to include non-confidential drawings and backup materials of the products and services in the Appendix.

Investors can believe in the worth of an IP while reviewing the business plan in case an entrepreneur is able to convince the prospective investor about how the IP is integrated into a product/service that yields real customer benefits in a large market. It is at this stage that an entrepreneur may discuss the matter regarding the signing of a non-disclosure agreement with the investor. While highlighting the importance of IP, an entrepreneur should do a good search through country-specific patent and trademark office Web sites, leading companies having IP strengths in the allied areas and the World Intellectual Property Organization (WIPO) to ensure the relative advantages that one's IP has over the existing IPs in the area. Some of the key aspects to be highlighted in the business-plan-related IP matters are the significance of IP owned in the success of ventures covering the following points:

- The extent of current use of one's IP assets and the potential benefits that can be derived.
- Type of IP assets and their relevance to one's venture.
- Does one's venture greatly depend for its commercial success on IP assets? If yes, then is the IP owned or licensed?
- While using IP, does one also have new products or processes that will provide a unique competitive advantage in the market? Are these going to give a turning point to the industry? How long does it take to derive such an advantage from IP?
- What are the key competitive advantages that one's IP assets would provide to one's venture?
- Explain how IP adds value to one's customers and contributes to developing a sustainable competitive edge.
- Mention about a plan, if any, that one has for gathering competitive intelligence.
- Does one gather IP information/databases for obtaining competitive intelligence on one's competitors?
- Are there any IP-related barriers to entering one's competitor's market, if so, clearly identify those and work out one's strategies to attract customers from competitors.
- Does IP help one in protecting those aspects of one's business that determine one's business success?

10. Legal Considerations and Insurance Requirements

A business plan should visualize and clearly understand the implications of various legal requirements and compliance thereof for the smooth running of the business. Otherwise, at a later date, an entrepreneur may waste their productive time in dealing with the legal hassles created because of unawareness, thereby resulting in non-compliance of legal provisions applicable to the business. As a business owner, one must understand their legal obligations and prepare to take the required actions to guarantee that they remain compliant with the law, and therefore, by going above and beyond legal requirements to serve as a model.

The first and foremost aspect is the formation of a company that gives a legal entity to the business. The legal structure that an entrepreneur chooses will have significant long-term implications for the running and growth of the business. An entrepreneur should also be clear about whether the business one proposes to enter into requires any government license or approval; if so, a plan of action for the same should be indicated in the business plan. An entrepreneur assumes a range of important health-

and safety-related responsibilities that require due compliance of the related provisions of law. Any implications of tax structure to be complied with have to be duly taken care of while planning, as these are an important considerations when doing business anywhere. Potential investors should consider income tax, capital gains tax and goods and services/transaction taxes in the region and industry of operations.

A business plan must also anticipate the effects of employment, property acquisition, environmental, foreign investment, and industry-specific regulations on the business, and prepare for taking the necessary actions at the right time to ensure the smooth execution of the plan.

A business plan should also take care of necessary insurances required such as employers liability insurance, public liability, and professional indemnity and insurance to protect the fixed assets of the company. Public liability and professional Indemnity may not be a legal requirement but help in protecting the business from compensation claims, especially if something goes wrong.

11. Management Team and Organizational Structure

A management team with complimentary skills, competency and knowledge is the most crucial aspect of a business plan that investors look for more minutely. The management team makes or breaks a business. The success of implantation of a business plan depends on the management team with a clear demarcation of roles and responsibilities, and the same clear understanding about the interdependency in the roles and responsibilities,

The Management Team section should include the profiles of promoters, biographies of the company's advisory board and/or board of directors, and key executives who matter most in the implementation of a plan. The presence of well-known advisors/ board members adds a lot of credibility to the business plan. It is important to describe how the presence of these advisors will directly impact the company through strategic advice and/or providing networking with key clients, partners and suppliers.

In short, this section helps in establishing and generating confidence in investors about the avail- ability of necessary talent to succeed. This section should be used to explain precisely how the team is uniquely qualified and experienced to effectively execute the venture in its present state. The details of some of the prominent members of the management team can be placed in the Appendix.

The section should cover the details of the team in such a manner that it reinforces a lot of confidence in the minds of investors that the management team knows its market, products and has the ability to successfully implement the proposed plan. In short, the business plan should be able to communicate the management's capabilities synchronized with the objectives laid down in the plan for the business In case there are deficiencies in one's team, particularly in a critical area, the entrepreneur should clearly highlight the strategy to fill up the voids at the appropriate time.

The essential elements of the management team that should be emphasized in the business are outlined as follows:

- Brief about promoters, board members, advisors and key executives while giving salient aspects of their resume vis-à-vis roles and responsibilities, hierarchy and decision-making process and projected hiring.
- Organizational structure highlighting the hierarchy, staff and line functions, key roles and people.
- Management team section should go over skill sets and track records of people. This section also presents details about key employees, pointing out qualifications, experience or outstanding skills; available competencies, knowledge and a skill set that will add a competitive edge to the image of the business. A mention is also to be made about special knowledge of budget control, personnel manage- ment, public relations and strategic planning.
- In the personnel plan, one's personnel structure and long-term plans for staff and their financial implications are described.

Thus, an organizational plan outlines the type of ownership along with an organizational chart, showing the hierarchy of authority and the obligations of the organization's members, as well as details about the promoter team regarding who will be taking a risk to support the project in practice. In short, a business plan should prove why a particular team is an outstanding one by duly highlighting the complimentary competencies, past track records and specific skills possessed that are essential for the success of the business plan, and achievements in terms of winning contests. It is essential to emphasize whether the team had collaborated previously and the accomplishments it had in terms of record sales, the launch of new products, and any significant breakthroughs, if applicable. A team member should be described 'generically' by highlighting their past credentials. For example, Mr X... has been employed by a major semiconductor manufacturer for the past 6 years.... He is presently serving as President, Innovation, Research & Development, with a team of 500 professionals reporting to him. His group has contributed 100 patents on an average for every six months resulting in introduction of innovations that are being continuously getting commercialized by the company for the past five years resulting in making them a leader in the semiconductor industry."

12. Risks, Problems and Assumptions

The development of a business is associated with inherent risks. Every plan invariably contains some implicit assumptions about various issues such as sales, cost, raising money, hiring people, developing products and a timeframe for the same. An entrepreneur should identify the various associated risks in the business and the implications and associated cost of adverse outcomes relating to one's industry, one's company and its people, a product's market appeal, and the timing and financing of one's start-up It is important to clearly define the critical assumptions concerning sales projections, customer orders, the cost of raw materials, the cost of raising funds and so forth. A business plan should comprehend risks and the strategies to manage or combat those risks. In case an investor identifies some critical risk factors that do not duly get taken care of by an entrepreneur, this may jeopardize the chances of raising funds. It should be remembered that after going through the Executive Summary, observing the management team and product/service characteristics, every investor looks at the risk factors to decide about their association with the venture or otherwise.

Major dangers for a start-up business may arise from the competitors' reactions and responses, negative fluctuations in sales stemming from market reaction, production-related challenges particularly regarding efficiency and quality, consequences of emerging technologies in the sector and insufficient understanding and clarity within the management team that may lead to a new product or service being rejected by the market. Therefore, various risk factors need to be properly analysed, and possible course of actions to manage them should be well comprehended in the business plan.

It has been rightly highlighted by John Mullins that 'the most common type of business plan, and the one that goes most quickly into the trash, is the one in which the writer can't find anything but good things to say about the opportunity and plans to pursue it'

Experienced entrepreneurs know better than to assert that everything is wonderful about their opportunity. They know there are potential pitfalls in their market or industry. The facts are that most opportunities are highly uncertain. Most new ventures will fail. Of the few that do succeed-winning capital, customers and positive cash flow it's usually not because of the original plan, 'Plan A', about which the business plan is written, but because of an as-yet-unknown 'Plan B'.

13. Launching a Business and Measuring Plan Progress

Having planned various aspects of the business to ensure its success, the real test lies in conceiving strategies to launch a business systematically without any tolerance for inefficiency or ineffectiveness. It has been said that the first fortnight and first 100 days of business unfoldment should be undertaken very carefully with meticulous planning. All those milestones that have affected the business at various times should be distinctly emphasized, and the method to accomplish this should be recognized. Usually, a good plan having pitfalls in launching can yield to failure.

Launching a business requires a strong management team coupled with adequate capital inflow at the appropriate time. Poor management has been a major cause of business failures and, therefore, should fetch a good deal of attention in developing a team that can successfully launch the business. The team has to be solution focused and has to handle all challenges on the way very effectively. A system needs to be devised and placed for checks and balances to regularly study, organize, plan and control all the activities of its operations. This includes the continuing study of market research and customer data, an area that is most sensitive for the success of the business.

An entrepreneur should chalk out various steps in sequence, highlighting the importance to be undertaken for smooth unfoldment of business. An entrepreneur should clearly highlight the points at which certain vital decisions have to be taken so as to ensure the progression towards goals and objectives is moving as scheduled from time and resources point of view. The entrepreneur must create a management information system for oversight and regulation that is utilized on a daily, weekly, biweekly, monthly, quarterly, semiannual, and annual basis. They should identify the items from big lists such as profit and loss statements, cash flows, inventory management, production, quality. sales, receivables, payables, urgency of payment and manpower to be monitored vis-à-vis projections/ budget. They should enable key team members to get feedback on time, so as to enable them to take corrective steps. The whole purpose to monitor the progress in implementation is to continuously update the plan on vital assumptions and measure the implications thereof. It is important to remember that an excellent business plan can become out of date with a crucial change in the conditions that were not visualized. The changes in a plan are inevitable on account of external factors, such as economic, market and technological developments, competition, political changes, and internal factors, such as key employee turnover and dissention among promoters. Therefore, ventures need to be sensitive to industry, company and market changes that may exert a major impact on the business plan assumptions.

14. Growth and Exit Strategy

Many start-up ventures get lost in the growth phase of the venture when opportunities are before them and they lack the required preparation to avail them. Therefore, a business plan needs to visualize the growth prospects of the existing venture or how new introductions through innovations would be continuously made on a growth path. The venture team should have a clear vision about the scalability of the venture and the emerging challenges that the growth would pose, so as to plan in advance the strategies for growth. The imminent challenge that the growth phase poses is decentralization, getting the right trustworthy people, focusing on systems, procedures and processes as against people.

The last portion of the plan has to explicitly state the exit strategy of one's business. Good business plans have inbuilt provisions for exiting from the existing business for personal, financial or opportunistic reasons. The most successful business plans develop contingency plans for multiple situations and enable promoters and investors to realize returns along with allowing the business to grow and expand.

Although at times it looks awkward to think of an exit, investors and, at times, entrepreneurs themselves would like to exit at the right time by either substantially gaining in the process or losing bare minimum. This requires long-term planning with a clear idea about the expected valuation of one's business as it grows in future. The entrepreneur should be very clear about the exit plan, as it has a direct bearing on the way the business is run.

For example, if the entrepreneur has decided to get listed on stock exchanges within say five to seven years of starting the business, they will have to necessarily think of introducing standard accounting practices accepted internationally from day one onwards. Similarly, if the entrepreneur is clear that they would like to pass over the business to the next generation at the earliest, they will be required to induct their children and train them well sufficiently in advance to have a smooth transition. The entrepreneur may decide in advance to hand over the business to professionals at a particular stage of growth when they would like to concentrate their energies on new businesses. This would require them to identify the right professionals who can be fully relied upon.

If one plans to pass on the business to one's children, one will need to start training them at a certain point.

At times, in the long run, the entrepreneur would like to exit by selling their shareholding, inducting investors and diluting their stake in the business or through liquidation. Instead, the entrepreneur may plan their exit strategy by going to the public, mergers with existing strong players or by selling outrightly.

15. Appendices

The details to support and reinforce the credibility of the assumptions made in the business plan are incorporated in the appendix, which is the last string in the whole plan. It contains certain confidential information that could be useful to stakeholders to strengthen their confidence in the business plan. Some of the vital information that is incorporated in the appendix is the resume of key managers, design and pictures of products, professional references to build credibility, market research study data sources and outcomes; patents; significant contracts lease, sales, purchase partnership, deed and insurance; product liability; director's liability, general liability; financial projections and assumptions; operation schedule and organizational charts.

The key purpose of items and information given in the appendix is to enhance the credibility of the business plan in the eyes of different stakeholders and, in particular, investors.

In summary, a business plan must be capable of emphasizing the main strengths like 'urgency' to seize an opportunity that might not present itself again or could diminish over time; distinctly demonstrate entrepreneurial capabilities and expertise in the field, rely on logic, reasoning, data and statistics, and, most importantly, exhibit a thorough comprehension of the actual environment—both present and forthcoming.

What is Most Important In A Plan?

As such, the most crucial and important aspects of a business plan depend on the nature of the business opportunity, but usually, it is the opportunity team, cash analysis and specific implementation details.

- Cash flow is the heart of a business-plan that is vital for the existence, sustenance and growth of the business. It is very hard to be followed in reality, as it depends on the product and customers' willingness to pay for the product upfront, in advance or at the stipulated time depending on the credit period. Cash is at times misunderstood as profits. Both cash and profits are two different things. It is necessary for a business to make profit, but earning profit without ensuring that the business generates sufficient cash at the time when it is needed will not ensure success. Since profits do not guarantee cash in the bank, lots of profitable companies go down the drain because of cash flow problems.
- Implementation acumen details of the management team are what make things happen. One's brilliant strategies and beautifully formatted planning documents are just paper exercise unless one assigns responsibilities, with dates and budgets, and follow up to those whom responsibilities are given to and

track results. Business plans are really about getting results as envisaged and improving one's venture on a continuous basis.

PITCHING A BUSINESS PLAN

Preparing a business plan is a great art and science to develop a clear path for the unfoldment of a business. However, it is not enough to prepare a good plan, what matters most is having clarity about the whole plan and how best the promoter team can present it to various stakeholders and, in particular, to investors for raising required funds. Pitching a plan effectively implies that the plan authors and the management team should be one and the same, even if some or all aspects of translating the vision of an entrepreneur into writing has been 'outsourced". Not only is the content of the business plan thoroughly examined by the various stakeholders, including investors, but the capabilities of the management team are also assessed based on their ability to deliver a presentation in a clear, concise and convincing manner. Investors take a decision to commit their funds on the basis of the capability of the management to execute and implement the business. Their main concern is not the document but an idea and the team that proposes to deliver the idea. There are different time slots for pitching an idea that have their own relevance, such as pitching in

- 10 seconds focuses on gut feel about the opportunity.
- 100 seconds is called 'elevator pitch'.
- 10 minutes focuses on gist of the presentation.
- 20 to 25 minutes deals with concise presentation.
- 45 minutes depicts full presentation.
- 90 minutes presents a whole story with detailed questions and answers.

Elevator pitch is the shortest possible time in which one needs to catch the attention of investors for generating further interest in finding out more details. Elevator pitch is a 60-second process that is a pretty short piece of information which people expect to contain some significant foundations, so as to capture the imagination of listeners and make them attracted to the business proposition. An entrepreneur ought to convey their business proposal, the reasons it would thrive in the market, and the type of financial backing they seek from investors. It is essential to possess a distinct strategy to articulate one's ideas without confusion and to employ impactful language and imagery; utilizing a few slides may be suggested in conjunction with the expression of ideas.

"Elevator pitch" means quick and clear description of a business that one could do in the time that one gets to share one's thoughts with a stranger in an elevator. This term is becoming more and more popular among entrepreneurs, venture capitalists and the teaching of entrepreneurship. Great skill needs to be inculcated by an entrepreneur. The whole essence lies in being clear, precise, simple, focused and powerful, so as to get easily bonded with prospective investors. It is also said to be the precise condensation at the heart of the business plan.

One should start one's experience in the elevator with a business situation and by narrating a story or incident; for example, personalize one's presentation by clearly identifying with the audience. For example, a business plan on biomass energy may highlight the increasing pressure and constraint on the existing non-renewable sources of energy that would pose a great threat to human existence and development, if not handled timely by the generation of energy through alternate sources of energy, more so renewable.

In the next part of your elevator pitch highlight with the strength of your solution to the problem of energy crisis, Why one? Why one's business? What is special about one that makes my offering or solution interesting to solve the problem for the identified target market? At this stage, the entrepreneur needs to bring into the picture the background, core competence and track record of his management team. For example, X, a certain co-founder, invented the best possible solution to a problem that has been proved well in terms of technical and commercial viability as evident from the patents in favour of the promoters and recognition got from business plan competitions organized by leading organizations such as Founder X being one of the most renowned experts in the field having international acclaims from leading energy organizations. Financial expertise lies with Y, who is a co-founder having experience in raising funds and managing fund uses for 15 years with leading orga- nizations such as.... One focuses on the competence of the team and the key differentiating factors that will result in the success of the venture. Further, one should briefly cover the money requirement and purpose there for, valuation of the venture right now, expected returns and exit strategy. One should be prepared with the help of facts and figures to logically answer the questions that would emanate from a presentation. What should be remembered is that the facts have to be presented with a power of words that catches fire in the minds of investors to get further interested in the opportunity and that has to be done quickly.

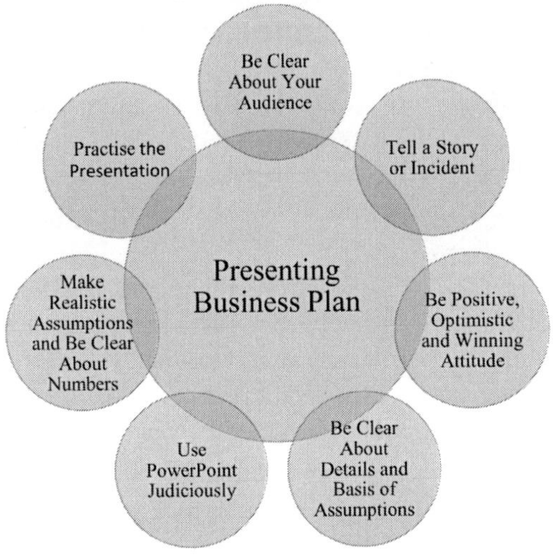

Figure 9.5: Key Tips to Effectively Present a Business Plan

If an entrepreneur is able to convince their stakeholders through an elevator pitch, they get a chance to present their idea in 15 to 20 minutes time. While presenting a detailed plan, one should remember the following key tips (Fig. 9.5)

- **Be Clear About Who Your Audience Are**-Knowing and engaging with audiences at a personal level by developing a rapport is a great art. Internet resources can be optimally and wisely used to research more effectively as also to develop a presentation targeted to a specific audience. Keeping in view the investors' motives and understanding their background are important to develop an effective presenta- tion. Pitch is used for various audiences such as:
    - Investors
    - key hires/co-founders
    - creditors/suppliers
    - possible partner companies
    - initial customers

  The emphasis and style of pitching has to get adapted to the target audience. Each audience has different expectations as to what is stored for them in the business plan. Therefore, they would be keen to look for their interest.
- **Tell a Story or Incident**-Placing an investment opportunity in context to a story or incident is very helpful in reinforcing the importance of a solution to the problem. In case an investor is able to relate well with the problem and proposed solution, then the chances of their association by way of investment become high. An entrepreneur must be capable of emphasizing the scale of the market potential along with the related profits that could be realized by seizing opportunities promptly.
- **Be Positive, Optimistic and Possess a Winning Attitude While Preparing**-Pitching to an investor is an outcome of hard work put in for weeks, months and, at times, years of planning. Mostly, for investors, it is the approach of an entrepreneur that focuses on their winning attitude and optimism coupled with great clarity about the execution of business, as reflected through the presentation of an executive summary that results in investors going through the full business plan.

  However, one should not exaggerate things to unnecessarily attract attention. For example, it is wrong to say that 'market is catching a wildfire and is bound to grow multifold". However, it may be right to say that 'most analysts agree with the market research outcomes vis-à-vis available studies that the market will grow at an average 40 per cent per annum for the next five years'.

  While presenting a team, it is not right to say, 'Now I would like to introduce our excellent Chief Executive Officer (CEO) who's going to rock your world. However, it would be appropriate to say. "We're really excited and fortunate to have a CEO who was a Fulbright Scholar, and has just sold his last company for $10 billion."

- **Be Clear About Details and Basis of Assumptions Made** - Investors desire a noticeable alignment between the presentation and the plan. Therefore, investors would pose direct questions to ensure that the entrepreneur has details with them and does not lend themselves into contradictions. Without fumbling in any manner whatsoever, the entrepreneur should be able to support their assumptions made with facts and figures during the presentation.

  One should refrain from making any baseless assertions in one's presentation. For example, 'our drug will get USFADA approval soon' is wrong, whereas if the presenter is sure, then they should say, 'we're well in advance stage of getting USFADA approval, as such our team of scientists have last 30 years experience in getting drugs approved, which sounds well.

- **Use PowerPoint Judiciously**-Although a PowerPoint presentation offers a great deal of visual appeal and simplifies the creation of an effective presentation, one must be very cautious not to rely excessively on it and simultaneously use it wisely. The quantity of slides should be limited to the essential minimum, approximately 15 to 20 for a half-hour presentation, with minimal content on each slide. The content should be rigorously analysed to ensure that relevance, clarity and time are managed carefully.

- **Make Realistic Assumptions and Be Clear About Numbers** - Investors minutely analyse and would like to understand numbers, production, market potential, funds required and return on investment. Therefore, all facts should be accurate and realistic. An entrepreneur presenting the plan should be well prepared to respond to the questions related to financial projections. Demand projections made for a product/service should be justifiable with facts and figures. It is important to realize that investors have alternate choices to invest their funds, therefore, an entrepreneur needs to sell their business plan effectively and prove superiority over other options that investors may have been simultaneously evaluating.

- **Practise the Presentation** - Overconfidence in having prepared and understood well does not serve the purpose. It resembles a proficient educator who excels in the classroom even after teaching the same subject for the nth time and is needed to invest the same amount of time (let's say two to three hours) or even longer to genuinely engage and spark students' interest in the topic at hand. Previous preparation by way of mock presentations or inside the team helps a lot. It may be desirable to submit an entry in a leading business plan competition that also provides a good opportunity to present the plan. An entrepreneur should seek feedback and review their earlier presentations to improve on a continuous basis. A credible business strategy and the ability to compellingly convey it significantly help in obtaining financial support.

  In short, pitching and presenting has a specific purpose of selling the idea to the investors and other stakeholders. The ultimate purpose is to convince investors and get their favourable response. Further, if one succeeds in getting their consent for investment, then the entrepreneur should have a full understanding about the terms offered vis-à-vis other options available and the knack of negotiating the terms well for the good of the business.

Keeping in view the aspects just referred to, an executive pitch deck lasting for around 20 to 25 minutes should focus on the following aspects in a presentation of 15 to 20 slides:

- Should clearly articulate vision and create excitement
- Should have three sections
- First two or three slides: executive summary highlighting pain point, unique solution to the pain point and size of the opportunity
- Approximately 10 more slides: details and further discussion
- Appendix: supporting research and assumptions made
- Be prepared to jump around, change order
- Highlight team constitution
- Include a slide on:
- Technology, market, business model, finance, marketing, competition, tactical/strategic milestones with timeline and risks

Dos and Don'ts for Pitching

Dos and Don'ts while pitching one's business plan before different stakeholders are given in the next few points:

- Involve all people from the team in the presentation by assigning appropriate roles. Be clear about one's audiences and what they are looking at in one's plan.
- Keep a hard copy of one's presentation, but do not hand it over beforehand or send it in advance. If required, while presenting, one can part with a hard copy with the audience.
- Be prepared to face audio-visual problems. Keep an alternate course of action ready rather than getting panicky about why technology is not working.
- Practice several times utilizing the computer that one will probably use during one's presentation.
- It is not at all advisable to read from the slides, so ensure that there is minimal text on them. Multi-media presentations can be used, provided they really add value to the overall presentation.
- Number the pages in the plan as well as the hard copy of the presentation.
- Use the executive summary rather than loading the presentation with big tables.
- Make the graphics crisp and reflective.
- Never ever have overconfidence. Rehearse with friends a number of times before one's final presentation.
- Be alert and aware of the content on the screen while responding to questions and answers.

- Even when reaching the conclusion of the presentation, it remains a significant challenge to engage audiences. Demonstrate total dedication and participation in one's plan while delivering the presentation.
- Thoughtfully incorporate humour, if it enhances the overall presentation. Employ humour or attempt to break through the invisible barrier, but exercise caution, and avoid resembling a used-car salesperson.

Reasons for Failure of Business Plans

The collapse of a business is significantly linked to inadequately researched and crafted business plans. Additionally, a lack of clarity regarding implementation specifics and the necessary team can lead to substantial setbacks, even when the opportunity is promising. Below are some frequently noted reasons for the failure of business plans:

- **Lack of Well-defined Vision**-A leader's vision is a scripted understanding of what a company wants to do and how they want to accomplish it. This provides a common purpose to all members of the organization to work for and focus their energies on a singular goal. Consequently, a poorly thought-out vision or the lack of a vision can lead to the downfall of businesses, since individuals operate at cross purposes when there is no unified direction.
- **Unreasonable and Unpractical Goals** - Being over-enthusiastic to set unreasonable targets not backed up by proper resources and market analysis can cause the failure of a business. It is important for start-up ventures to be reasonable and get a taste of success gradually with small steps to be climbed in the beginning to gain experience as also ensure that assumptions made are more or less working out to be correct.
- **Lack of Quantification of Goals and Objectives** - It is crucial to measure objectives in relation to sales, profits, costs, and the physical resources needed in terms of the number of employees at various levels and raw materials. The intangibles also need to be quantified using specific techniques and by creating suitable indicators. For instance, customer satisfaction can be evaluated based on the quantity of complaints, the time and resources dedicated to resolving complaints compared to the total number of customers utilizing company products/services.
- **Starting a Business for the Wrong Reasons**-At times, plunging into an entrepreneurial journey with the wrong reasons such as making a lot of money, getting a lot of time to spend with the family and not being accountable or answerable to anybody results in a business failure. One must be very clear in one's mind regarding the reasons and rationale for engaging in the business and its implications from both a professional and personal life perspective.
- **Poor Management and Lack of Commitment** - If leadership lacks the aggressive and 'driven ambition' to succeed, it can quickly jeopardize the business. It is crucial to have enthusiastic and dynamic leaders who will stop at nothing to secure the support, resources, and technology necessary. Numerous studies have pointed to inadequate management as the primary cause

of business failures. Particularly, start-up teams often lack essential business and management knowledge in key areas such as finance, purchasing, selling, production, and hiring and managing staff. In addition, not being fully engaged and dedicated to the business can lead to its collapse. Proper attention should be devoted to routinely studying and analyzing, organizing, planning, and controlling various operational activities in order of their importance.

- **Inadequate Capital**-A common mistake that kills many businesses is having insufficient operating funds. Business owners often undervalue their financial needs and find themselves having to shut down before they have had a reasonable opportunity to thrive. Remember, usually in the midst of an entrepreneurial journey when one is badly in need of fund support, one will not get it. One should also not have an the unrealistic expectation of incoming revenues from sales and find themselves in trouble to meet commitments because of non-receipt or delayed receipt of sales revenue from customers.

  Therefore, it is extremely crucial to gauge with a higher level of confidence the need for financing for the business at various intervals concerning the expenses of initiating and operating a business as well as the expenses associated with remaining in the business. It is essential to factor in that numerous businesses require a more extended timeframe to achieve a breakeven point, thereby indicating the necessity for sufficient funds to address all expenses until revenue can ultimately cover all production costs and yield profits.
- **Lack of Relevant Experience and Knowledge of the Business** - At times, insufficient knowledge of the business market and pertinent experience presents a significant risk to the survival of the business, unless these deficiencies are professionally addressed by the introduction of suitable personnel. It is essential that the core management team conducts comprehensive research to acquire a profound understanding of the elements that are crucial for a business's success. A lack of adequate understanding and insight regarding the business can result in the failure of the company.
- **Government Regulatory Measures** - The owner of a business should have a good grasp over the rules and regulations governing their sector of the economy, this includes having a clear understanding of the tax structure and other statutory compliances. Otherwise, the entrepreneur gets into unwanted legal hassles and gets forced to waste a lot of their productive time in sorting out unproductive legal and regulatory issues. Not knowing the extent of government interference and the applicability of regulatory mechanisms can mean the difference between the success and failure of a business.
- **Lack of Strategies and Preparedness to Respond to Potential Threats** - Business ventures that get caught because external threats emanating from change in government policies, natural calamities, technological developments and introduction of better-quality and lower-cost substitutes by competitors and that are not prepared to respond to these challenges by having intrinsic

strengths or innovative strategies to revitalize the business model may face a setback. Entrepreneurs with a long-term insight will always keep resources for facing such eventualities and respond to such challenges, so as to come up with greater strengths.

- **Lack of Clarity About Target Market and Customer Needs** - The success or failure of a business greatly depends on better clarity about market and customer needs. Understanding the market and customer needs well through deeper analysis and research paves the way for offering what the market and customer want and would be willing to pay what the entrepreneur expects. New ventures, at times, lend themselves into a great problem by offering something that the market is not prepared to accept at a given point of time. This gives rise to a big jolt to the existence of the venture.
- **Unreliable Suppliers**-The delivery of a product/service on time to the customer creates a lot of good- will and brand building. The customer always looks forward to receiving the product before or exactly at the time committed by the entrepreneur. To ensure timely delivery, the entrepreneur should tie-up for buying resources from the suppliers who would deliver to them on time. Any dependency on suppliers who are not likely to meet their deadlines can cause a lot of problem to the entrepreneur. Therefore, the entrepreneur should only put up orders and develop relationships with the suppliers who would not only supply on time but also help entrepreneurs in meeting sudden valuable commitments. It is rightly said that "You can't sell what you don't have. Your ability to maintain proper levels of inventory matters the most "in success of a business". Maintaining required inventory depends on the quality of entrepreneurs' relationships with reliable suppliers. Developing effective supply channels does take time and investment by way of efforts. In case an entrepreneur has any problem with the supplier, it is advisable not to drag things but to take action! Look for new dependable and reliable supplier and switch over at the earliest.

Key Concepts

- **Planning:** Is a process that helps a lot in improving the chances of success and prepares well in advance to respond to the challenges that may emerge on the way.
- **Business Plan:** Is the formal written expression of the entrepreneurial vision, describing the strategy and operations of the proposed venture.
- **Business Plan Drivers:** Differentiate the business from their competitors by creating value vis-à-vis minimum possible cost.
- **Vision:** Defines the desired or intended future state of an organization or enterprise in terms of its fundamental objectives and/or strategic direction.
- **Perspective of the Entrepreneur:** The main concern of an entrepreneur is to successfully translate their ideas into a reality that would give rise to sustainable growth.

- **Marketing Perspective:** Has to clearly explain the process through which an entrepreneur will make their customers buy their products/services. This perspective has to clearly incorporate unique product/service propositions that one has and one's unique selling proposition.
- **Investor's Perspective:** Basically looks for the profit-generating capacity of an idea in the business plan.
- **Executive Summary:** Provides a precise and attractive gist and highlights the key points related to a product/service, funds required, returns expected, market potential, the team backing it and their commitment.
- **Competitive Analysis:** Captures the competitive landscape of a business by identifying direct and indirect competitors and assessing their strengths and weaknesses.
- **Market Research and Analysis:** Assess various customer segments that the venture would serve by clearly identifying the needs of the customers that a business venture proposes to satisfy.
- **Marketing Plan:** Develops concrete estimates for sales and the strategy that will enable the entrepreneur to avail the opportunity in the light of specific competitive advantages.
- **Financial Plan:** Assesses the financial implications of various strategies proposed in other sections to quantify flow of profit, cash vis-à-vis investments required.
- **Operational Plan:** Highlights key aspects such as product or service characteristics; production, planning and control aspects; location, equipment, people, processes and the surrounding environment.
- **Organizational Plan:** Describes form of ownership, organizational chart, indicating line of authority and responsibilities of members of the organization, and the details of the promoter team who would be taking a risk to back up the project in reality.
- **Exit Strategy:** Provides an inbuilt provision for exiting the existing business for personal, financial or opportunistic reasons.
- **Elevator Pitch:** Is the shortest possible time in which one needs to catch the attention of investors for generating further interest in finding out more details.

10

Industrial Sickness

In general, an unwell entity is one that fails to regularly create a surplus and frequently relies on outside funds to exist. The Sick Industrial Enterprises (Special Provision) Act 1985 characterized sickness as "a financial loss over two successive accounting periods and a total loss that is equal to or exceeds the net assets of the enterprise at the conclusion of the second accounting period".

Revised Definition of Sick Smallscale Industry "A small scale industrial unit is deemed sick if it has incurred losses at the conclusion of the accounting year that are equal to or exceed 50% of the highest level of its net assets in the immediately preceding five financial years".

Symptoms of Sickness

The continuation of many signals for a long time is called a symptom of a disease, and the various symptoms are ultimately reflected in the operation of the facility (capacity utilization rate, financial capital ratio, stock market value practice in various financial areas, production, marketing and labor relations in the industry).

Among the key signs are:

- Deterioration of financial relations
- Review delay
- Lack of constant cash flow
- Continued decline in stock prices
- late payment obligations
- Extensive use of creative accounting
- Repeated loan requests
- Deterioration of employee morale
- Desperation between senior and middle management.

However, in each case, the financial ratios cannot be regarded as accurate indicators of industrial disease.

Causes of Industrial Sickness

The following are causes of small unit health failure:

- Production issues

- A lack of orders
- A lack of raw materials
- A cash crunch
- An increase in raw material costs, and
- A few others

1. **Production Issues**: Production issues might emerge as a result of:
    - Machine breakdowns or insufficient maintenance
    - Low quality of raw material
    - Insufficient labor productivity
    - Power failure
    - Inadequate production, planning and control
    - Delays in subcontracted deliveries
    - Inadequate working conditions.
2. **Lack of Orders:** An organization may experience difficulties if it fails to obtain sufficient orders to achieve projected or established production targets. The shortfall in orders could be attributed to one of the ensuing reasons:
    - Competitive
    - Economic recession
    - Low quality technical failure
    - Irregular delivery
    - Ineffective marketing
    - Expiration
3. **Lack of Raw Materials:** A lack of raw materials could be caused by:
    - National or regional deficit
    - High price
    - Late payments to suppliers
    - Poor quality
    - Uncertain suppliers
    - Lack of preparation
4. **Cash Crunch:** This can occur as a result from:
    - Deliberate but irrational dispersion of assets
    - Intentional but irrational diversification (e.g. unintentional diversification)
    - Bad collections
    - Unscheduled payments to creditors
    - Excess inventory

- Unnecessary and pointless expenses

5. **Increased Raw Material Cost:** This is a common cause that might lead to a unit's sickness. It could be due to:
    - A rise in expenses that are not shown in sales prices due to inaccurate cost accounting
    - In an inflationary market, large orders are recorded at a fixed price
    - Serious material loss
    - Expensive warehouse
6. **Increased Overhead Costs:** Increased overhead expenses may result from:
    - Inefficient production
    - Underutilization of capacity
    - Large loans and excessive interest charges
    - Increase in administrative or sales expenses
    - Unplanned investments
    - Development or diversification of new products

Sickness Prevention

Disease units have been studied extensively. Some units are born sick, some units are created sick, and some units are sick. Four classifications can significantly contribute to maintaining units in good condition.

These are as follows:

- Term loan institutions (lending institutions)
- Commercial banks
- Entrepreneurs
- Board of Directors

1. **Term Lending Institutions (loan institutions):** Sickness can be avoided if term loan institutions take the following steps:
    - Appropriate assessment of project management.
    - Determining the marketability of the final products of the unit before granting the loan
    - Special care in examining the acceptability and technical feasibility of the equipment and machinery used
    - Timely completion of the project is ensured
2. **Commercial Banks:** As commercial banks provide the necessary facilities for successful service, they must take the following precautions to prevent diseases:
    - Please note that withdrawals from the account do not exceed the permitted withdrawals

- Regular monitoring of the ledger account with master data
- Frequent communication with the contractor

3. **The Entrepreneur:** The ability and determination of the entrepreneur are undoubtedly important in the successful operation of the unit. The entrepreneur creates, implements, and monitors the project to keep the entity healthy.

   They can avoid illness by:

   - Careful selection of the machine supplier
   - Do homework
   - Great attention is given to the shortcomings of the unit
   - Careful selection of partners

4. **The Government:** The government can also help curb the disease by not making sudden changes in industrial policy, such as the following, which can destabilize small units

   - Multiple changes to existing units
   - Restriction of new investments
   - Abolition of subsidies
   - Open reserved websites to large companies

11

Sources of Finance

1. National Sources of Finance

A. Medium and Long-term Sources

i. **Venture Capital:** Funding from affluent investors who prefer to invest their money in enterprises with long-term growth prospects is referred to as venture capital. The individuals who provide such investments are known as venture capitalists. Venture capitalists usually acquire equity in the emerging company in return for their backing, which is typically provided as preferred stock.

Venture Capital is a funding source for new businesses. 'Venture capital refers to investments made in 'new and untested enterprises' that do not have a solid growth history'. It comes as capital invested as shareholdings, aimed at establishing and creating firms focused on innovative concepts or new technologies that carry a high perceived risk, yet possess the potential for rapid expansion. Venture capital is generally arranged as equity or quasi-equity financial instruments. A venture capital firm acts as an intermediary between investors seeking high returns on their investments and entrepreneurs who require funding to launch their businesses.

Types of Venture Capital Funding

The various types of venture capital are classified according to their applications at various stages of a company's lifecycle. Early stage financing, expansion financing, and acquisition / buy out financing are the three main types of venture capital.

The venture capital funding procedure is completed in six financing stages that correspond to the stages of a company's development. These stages are:

- **Seed money:** Low-level financing used to test and grow a new idea.
- **Start-up:** Funding for new businesses to cover marketing and product development expenses.
- **First-Round:** Manufacturing and early sales funding.
- **Second-Round:** Operational capital is provided to early-stage companies that sell products but do not make a profit.
- **Third Round:** Also referred to as Mezzanine financing, this is capital utilized to grow a recently successful business.

- **Fourth-Round:** Also known as bridge financing, the fourth round is proposed for financing the "going public" process.

ii. **Seed Capital:** Seed Capital can be described as

- It is an initial contribution by the owner
- It is returnable after the function is over
- It is provided by the owner to allow the initiation of preliminary activities of the Project.
- It enables to commence the initial activities

iii. **Equity Share Capital:** It is the original finance supplied by the company's promoters for the project. It is the enduring long-term capital contributed by the owners/promoters – whether Indian or Foreign in the case of a Joint Venture Company. Equity capital is also contributed by the Public and by investment institutions like LIC, financial institutions, UTI, etc. Investments in equity capital are regulated by the rules of SEBI (Securities Exchange Board of India).

iv. **Preference Share Capital:** Initial capital for a project can be raised through the issuance of preference shares. The difference between the preference shares and equity shares is that preference shareholders get a preference over the equity shareholders in the event of liquidation. Preference shareholder gets a fixed dividend unlike a Equity shareholders, whether profit or no profit earned by the company.

v. **Debentures or Convertible Debentures:** Funds are also raised by the issuing a Debentures. Debentures have a set interest rate that can be redeemed after a specified time period. At the end of stipulated time, the issuing company buys the debenture back or converts them into equity shares or pays back.

vi. **Leasing:** Company can also obtained fixed assets from leasing companies on agreed terms and conditions. The leasing companies or leasing financial institutions acquire the fixed assets. Terms and condition agreed by the both for the compensation towards the fixed assets leased by the lesser.

vii. **Bonds:** Funds for projects are also raised by the Bonds. Issue of Bonds is to be sanctioned by the Ministry of Finance, Rate of interest on bonds must be approved by the Government. These bonds can be repurchased by the issuing company following the completion of the specified time according to the regulations.

viii. **Public Deposits:** It serves as an effective means of generating funds for projects. It must be governed in accordance with the stipulations set forth in the Companies Act.

ix. **Institutional Finance Sources:** Financial institutions serve as the most effective sources for substantial funding for major projects over extended periods, as well as during business activities for modernization, replacements, and repayment of long-term loans, among other needs.

Financial Institutions adhere to a stringent procedure of assessment and evaluation when offering funding for the projects.

Lists of financial institutions are given below

- Life Insurance Corporation of India (LIC)
- Unit Trust of India (UTI)
- Industrial Finance Corporation of India Limited (IFCI)
- National Industrial Development Corporation (NIDC)
- Industrial Development Bank of India (IDBI)
- Small Scale Industrial Development Bank of India (SIDBI)
- State Finance Corporation

LIC and UTI generally invest in the shares of the companies, within the rules and powers delegated to them.

B. Short Term Sources

i. **Bank Loans:** Loan from commercial bank includes bank loan, cash Credit, overdraft besides various other financing facilities from the bank. These banks provide financing in the form of
   - Cash Credit
   - Short-term Loans
   - Long-term Loans
   - Bills Discounting
   - Bank Guarantee
   - Letters of Credit

   For short-term financing, the interest rate is elevated; thus, the funds acquired from banks via cash credits and short-term loans are primarily used to finance working capital. In contrast, for a long-term loan, the interest rate is relatively low, and it is utilized for funding a project. However, the availability of long-term loans from commercial banks is restricted.

ii. **Trade Credit:** It is between the two organizations for the commercial transactions on mutually agreed terms and conditions.

iii. **Government subsidies or Grant:** Government of India establishes dedicated funds for the advancement of specific projects. Occasionally, these initiatives are carried out with collaborative financing, meaning partially from the Government and partially by the beneficiary organization.

iv. **Deferred Credit:** Project funding is accomplished through delaying the payments.

C. Internal Sources

1. The following are the different sources to generate funds internally for use within the organization.

- Internal Funds
- Retained Earnings
- Provisions
- Sale of Idle assets
- Dividend Policy and retention of Profits
- Bonus Share
- Deferred payment arrangements

2. **Sources of Finance (International):** Capital markets have broadened their scope with the emergence of global financiers like FDI, ECB, GDR, etc. Financing is also accessible through partnerships with foreign companies that include equity participation for technology transfers.
3. **Miscellaneous Sources of Finance**

Financial Institution

Institutional framework

To offer financial support to entrepreneurs, the Government has established several specialized financial institutions in addition to commercial banks. They can be categorized into two groups.

1. **State level Financial Institutions:** The State Level Institutions are categorized into two groups, namely State Financial Corporations (SFCs) and State Industrial and Investment Corporations. In addition to these two institution types, certain states also have state industrial corporations that focus on developing industrial estates and offering financial aid such as equity support.
2. **All India Financial Institutions:** The All India Level Finance Institutions are as follows:
   a) Industrial Finance Corporation of India
   b) Industrial Credit and Investment Corporation of India
   c) Industrial Development Bank of India
   d) Industrial Reconstruction Bank of India
   e) EXIM Bank of India
   f) Unit Trust of India
   g) LIC, GIC, etc.
   h) National Small Industries Corporation.
   i) Small Industries Development Bank of India.

A financial institution (FI) refers to a company involved in the activities of managing financial and monetary transactions like deposits, loans, investments, and currency exchange. Financial institutions cover a wide array of business activities within the financial services industry, which includes banks, trust companies, insurance companies, brokerage firms, and investment dealers.

Types of Financial Institutions

Financial establishments provide a broad spectrum of products and services for personal and business clients. The particular services available differ significantly among various categories of financial institutions.

1. **Commercial Banks:** A commercial bank is a kind of financial organization that takes deposits, provides checking account services, grants business, personal, and mortgage loans, and supplies essential financial products such as certificates of deposit (CDs) and savings accounts to individuals and small enterprises. A commercial bank is the place where the majority of individuals carry out their banking activities, in contrast to an investment bank.

   Banks and comparable financial institutions, like thrifts or credit unions, provide the most widely acknowledged and regularly utilized financial services: checking and savings accounts, home loans, and various forms of loans for individual and business clients. Additionally, banks serve as payment facilitators through credit cards, wire transfers, and currency exchange.

2. **Investment Banks:** Investment banks focus on delivering services aimed at supporting business activities, such as financing for capital expenditures and equity fundraising, which includes initial public offerings (IPOs). They frequently provide brokerage services for investors, serve as market makers for trading platforms, and oversee mergers, acquisitions, and various corporate reorganizations.

3. **Insurance Companies:** Among the most recognized non-bank financial entities are insurance firms. Offering insurance, be it for individuals or businesses, ranks as one of the earliest financial services. Safeguarding assets and shielding against financial risks, provided through insurance offerings, is a vital service that supports individual and corporate investments that drive economic development.

4. **Brokerage Firms:** Investment firms and brokerages, like the mutual fund and exchange-traded fund (ETF) provider Fidelity Investments, focus on offering investment services that encompass wealth management and financial advisory services. They also grant access to investment products that may vary from stocks and bonds to lesser-known alternative investments, such as hedge funds and private equity investments.

Finance from Banking Institutions

Institutional finance for large, medium, small, and tiny industries by commercial banks - the State Bank of India group, nationalized banks, private sector banks, and development corporations which have been specifically created to offer industrial finance. Additionally, the Reserve Bank of India provides credit guarantees and the ECGC offers export guarantees to the small-scale sector. Through its refinance activities, the Industrial Development Bank of India also contributes significantly to the development of the small scale-sector as it has allowed the SFCs, SSIDC/SSIACS, and commercial banks to provide substantial amounts of financial support to this sector.

IDBI and SIDBI are top banks that offer refinancing services to other institutions. Similarly, NABARD serves as a primary bank for agricultural financing and Exim Bank specializes in export-import transactions. Additionally, there are industrial development banks, specialized institutions, savings and investment institutions, financial service institutions, and regulatory bodies.

Commercial banks, State Financial Corporations, State Directorate of Industries, National Small Industries Corporation, state Small Industries Corporations, and all-India development banks provide financial support to entrepreneurs. The credit facilities provided by commercial banks and State Financial Corporations fall under the Credit Guarantee Scheme for Industries, which safeguards credit institutions against potential losses from their lending activities within this sector.

1. **Industrial Finance Corporation of India (IFCI):** The Industrial Finance Corporation of India (IFCI) was founded in 1948 through an Act of Parliament aimed at offering medium and long-term financing to industrial entities in India. IFCI evolved into a corporation on May 21, 1993, to deliver increased flexibility to respond to the demands of the swiftly evolving financial landscape.

   The Board of Directors comprises a full-time Chairman and twelve directors. The Chairman is designated by the Central Government following discussions with the IDBI. Two directors are appointed by the Central Government and four by the IDBI. Six directors are chosen by shareholders excluding the IDBI.

   Financial support offered by the IFCI can be in one or more of the subsequent forms:

   - Rupee and foreign currency term loans
   - Underwriting of share and debenture issues
   - Direct subscription to equity
   - Guarantees
   - Soft loans
   - Equipment financing Projects with expenses up to Rs. 300 lakh are funded by the State Financial Corporations, State Industrial Development Corporations, and Commercial banks through the refinance scheme of the IDBI. Assistance from the JFCI is only granted to projects that exceed a cost of Rs. 300 lakh.

2. **The Industrial Development Bank of India (IDBI):** The Industrial Development Bank of India (IDBI) was founded on July 1st, 1964 under the Industrial Development Bank of India Act, as a fully owned subsidiary of the Reserve Bank of India. According to the Public Financial Institutions Laws (Amendment) Act, 1975, the ownership of the IDBI was transferred to the central government, effective February 16, 1976. The IDBI's most unique attribute is that it has been designated the principal financial institution responsible for coordinating, in alignment with national priorities, the activities

of entities involved in financing, promoting, or developing industry. The IDBI has been given a specific role to fulfill regarding industrial development.

Objectives and Functions

- To function as a primary institution for term finance for the industry, to coordinate the operations of entities involved in financing, promoting, or developing industries, and to aid in the advancement of these entities.
- To strategize, promote, and enhance industries to address deficiencies in the country's industrial framework.
- To offer technical and administrative support for the promotion, management, or expansion of industries.
- To conduct market and investment research and surveys, as well as technical and economic studies related to industrial development.
- To serve as the lender of last resort and to finance all varieties of industrial enterprises engaged, or planning to engage, in the manufacture, processing, or preservation of goods, or in mining, shipping, transport, hotel businesses, or in power generation and distribution, in fishing or in providing shore fishing, or in the upkeep, repair, testing, or servicing of machinery or vehicles, vessels, etc. , or for establishing industrial estates.
- The Bank may also support industrial enterprises involved in the research and development of any process or product or in supplying special or technical knowledge or other services to foster industrial growth. Additionally, it provides financing for the export of engineering goods and services on a deferred payment basis.

  The IDBI has been significantly involved in promoting small-scale industries. Its support has been provided through its refinancing scheme for industrial loans and, to a minor degree, through the Bills Rediscounting Scheme.

3. **The Industrial Credit and Investment Corporation of India (ICICI):** The ICICI (Industrial Credit and Investment Corporation of India) was envisioned as a private sector development bank in 1955, primarily aimed at supplying development finance to the private sector. Its current goals encompass:
   - Helping in the establishment, growth, and updating of businesses;
   - Fostering and advocating for the involvement of private investment, both domestic and foreign, in the ownership of industrial investments and the growth of investment markets.

   Financial support is being offered by ICICI in these ways:
   - Rupee and foreign currency term loans
   - Underwriting of share and debenture offerings
   - Direct subscription to equity
   - Guarantees
   - Soft loans

- Suppliers' credit line to facilitate the sale of industrial equipment under deferred payment conditions
- Financing through leasing
- Financial Indo-US collaborations in research and development. In practice, only projects exceeding Rs. 300 lakhs are eligible for financial support from ICICI. However, there is no minimum project cost requirement for foreign currency loans.

4. **The National Bank for Agriculture and Rural Development 'NABARD':** The goals for creating a bank to be named the 'National Bank of Agriculture and Rural Development' aim to offer credit for the advancement of agriculture, small-scale industries, cottage and village industries, handicrafts and other rural crafts, as well as additional allied economic activities in rural regions to foster comprehensive rural development and ensure the well-being of rural areas, and for matters related or ancillary to this.
5. **The Small Industries Development Bank of India (SIDBI):** To guarantee an increased influx of financial and non-financial support to the small-scale industry, the Indian government established the Small Industries Development Bank of India (SIDBI) through a Special Act of Parliament in 1989, making it a wholly owned subsidiary of the Industrial Development Bank of India (IDBI). The SIDBI has assumed responsibility for the existing portfolio of the IDBI associated with the small-scale sector.

   The important functions of SIDBI are as follows:

   - To initiate steps for technological up gradation and modernization of existing units.
   - To expand the channels for marketing the products of SSI sector in domestic and international markets.
   - To promote employment-oriented industries especially in semi-urban areas to create more employment opportunities and there by checking migration of people to urban areas.
6. **Industrial Investment Bank of India:** The Industrial Investment Bank of India (IIBI) was founded in 1985. The range of IIBI's financing activities has expanded following the removal of the Government requirement that 60% of its portfolio be composed of struggling companies; IIBI now funds all industrial projects like any other financial institution.

   IIBI provides loans and advances to industrial enterprises, underwrites stocks, shares, bonds, and debentures, and offers guarantees for loans/deferred payments. It finances the acquisition of equipment and makes machinery and other equipment available on a lease or hire purchase basis. It also offers infrastructure facilities, consultancy, managerial, and merchant banking services. IIBI has planned the establishment of a Special Fund, known as the Reconstruction Assistance Fund, to address the unique financial needs of supported medium and large-scale units for their recovery and rehabilitation, which cannot be fulfilled by banks and financial institutions under typical conditions.

7. **Export-Import Bank of India (Exim Bank):** The Export Import Bank of India (Exim Bank) was established on January 1, 1982, through an Act of Parliament as the main financial body for supporting and financing India's international trade. Exim Bank provides financing to exporters and importers, coordinates the operations of institutions involved in financing the export and import of goods and services, funds export-oriented units, and carries out necessary promotional activities for international trade. It offers a range of 23 major programs to cater to the requirements of various customer segments. Exporters can access pre-shipment credit, suppliers' credit, and overseas investment financing; loans for export product development, export marketing loans, bulk import financing, and vendor development finance. Buyers' credit and lines of credit are made available to foreign governments and agencies. For commercial banks in India, Exim Bank provides the facility for rediscounting export bills, refinancing suppliers' credit, and refinancing term loans related to export-oriented units. It also engages in guarantees provided by commercial banks on behalf of Indian project exporters. In addition to financing, EXIM Bank enhances exports by offering advisory and information services to exporters regarding procurement practices and bidding processes of multilateral institutions, country risk assessments, merchant banking, and marketing aimed at boosting exports of non-traditional products to developed nations.

Estimating Financial Requirements

Before estimating how much money need for business, entrepreneurs must know what they plan to do, unfortunately, and many entrepreneurs do not, often because they have failed to work out business plans. The preparing proper plan enables entrepreneurs to crystallize their thinking on how best to launch their ventures. It forces them to move logically and systematically from the stage of dreams and ideas to that of concrete action. The concreteness of business plans helps entrepreneurs determine their financial needs.

The center piece of the business plan is the cash budget. Without a cash budget, the entrepreneur has no way of estimating financial needs. So vital are cash budgets that few investors or creditors will entertain a request for money without one. The cash budget enables them to decide intelligently whether to finance the entrepreneur.

Preparing a Budget: Let us now discuss the details of how to work up a cash budget. Before beginning to develop a cash budget, an entrepreneur must first spell out his or her operating plans, defining production, marketing, staffing, accounting, and legal goals. Take note that these are all essential components of the business plan.

Before we outline the procedure for converting these plans into rupees, let us initially highlight two constraints of budgeting:

- All budgets rely on predictions of future events. Nevertheless, no one can foresee the future with complete accuracy, thus budgets can only be as good as the predictions. Thus the entrepreneur should be as thorough as possible in his or her efforts to prepare workable operating plans.
- A budget cannot account for the effects of intangible qualities or unpredictable events. It cannot reflect how skilled and able the entrepreneur may be, nor can it reflect teamwork and morale. A budget can only address upcoming events that can be represented in rupees.
- Still, budgeting represents a remarkable achievement. It provides a way of summarizing the future in a single statement, in language that investors and creditors understand. Let us now see how operating plans may be translated into a cash budget:

Example: An entrepreneur intending to launch a home furnishings store has projected her sales income for the initial three years. This revenue forecast is a result of the marketing plan that she worked out as part of her business plan. Although it is challenging, the revenue projection is the most critical estimate an entrepreneur can establish, as it serves as the foundation for the majority of the other estimates. For instance, a retail business with annual revenues of $2 million instead of $400,000 might require five times the number of sales staff, four times the amount of retail space, and three times the level of inventory.

Having forecast her revenues, the entrepreneur next estimates the cost of the fixed and current assets she will need to support those revenues. Before we proceed with our example, we will describe and differentiate fixed and current assets.

Fixed Assets: Fixed assets are resources whose use will benefit the entrepreneur for more than one year. An example is a building bought for $150,000. If the entrepreneur expects the building to last 25 years, he or she would receive $6000 worth of shelter benefits a year. Other examples of fixed assets include machines, land, trucks, desks, and even ashtrays.

These examples are resources the entrepreneur can touch and see, but fixed assets may also be intangible. For example, an inventor may sell an entrepreneur the patent rights for a new pollution-control device for $80,000. The entrepreneur might expect to benefit from the patent rights for the next 10 years. These rights cannot be touched or seen, but they are a long-lived asset that will provide benefits to the entrepreneur's business for more than one year. Other examples of intangible assets are licenses and goodwill.

Current Assets: In contrast to fixed assets, current assets are resources whose benefits will last less than one year. Commonly, current assets are cash, accounts receivable, and inventories. Accounts receivable - the bills owed by customers who buy on credit - represent a current asset because the entrepreneur can expect to collect within a short time. Similarly, inventory is a current asset because the entrepreneur usually expects to recover the investment in inventory by selling it within a short period after purchasing it from a supplier. Let us now return to our previous example.

Example: The entrepreneur's store will sell Scandinavian furniture. In estimating start-up costs, she has decided to

- Construct a one-story, free standing building with 5,000 square feet of floor space to display and store furniture. Cost: $ 150,000 at year zero.
- Keep a base inventory of furniture large enough to generate twice the average monthly forecast revenues; in addition, buy enough inventories monthly to cover the following month's budgeted revenues. She plans to pay for all inventories within a month of purchasing it. Cost: $36,000 at year zero.
- Put an asphalt surface on a parking lot next to the building. Cost: $16,000 at year zero.
- Finance customers who buy on credit. The entrepreneur is assuming that all sales will be credit sales, with customers taking a month to pay, on the average. (One month is the industry's average collection period.)
- Buy fixtures, office equipment, and a half-ton delivery truck. Cost: $ 26,000 at year zero.
- Incorporate the venture with a lawyer's help. Cost: $ 2000.
  - Design a record-keeping system with the help of an accountant. Cost: $ 1,000.
  - Buy a three-year prepaid insurance policy. Cost: $6000.
  - Buy city, county, and state licenses. Cost: $ 1,000.
  - Promote the store's grand opening. Cost: $2,000.

So far, the entrepreneur has estimated what it would cost just to open for business. She now must go one step further and estimate what it would cost to stay open through the first year, by month:

Monthly Cash Expenses (excluding cost of goods sold):

Entrepreneur's salary $2,400

Part-time employee wages 1,200

Advertising 600

Electricity, heat, telephone 400

Delivery 400

Accounting, legal 400

Supplies 200

Other 800

Total $ 6,400

Note that these monthly expenses are unlikely to change with revenues. That is even if first-year revenues were double the $ 249,000 forecast, monthly expenses would not be significantly greater than $6,400. The only expense item likely to increase significantly would be part-time wages, since as revenues increase; the entrepreneur will probably need to add more part-time salespeople to wait on customers.

Working Capital Financing

Working capital is the lifeblood of a business. Its adequate planning and proper management is necessary for the successful operation and continued existence of a business. Efficient management of working capital is a basic necessity for sound operational health of a every enterprise. Therefore, working capital management is an integral part of business management. Working capital management basically means management of current assets, current liabilities, and interrelationship between the two. Working capital is of a liquid nature. Therefore, working capital management is also called 'liquidity management'.

Concept of Working Capital: Working capital is understood in the following two ways:

Net Working Capital: Net working capital refers to the surplus of current assets beyond current liabilities. To put it another way, it can be defined as the allocation of long-term (non-current) resources for current assets. The application of the net working capital concept is qualitative since it offers an assessment of a company's financial wellbeing. The higher the amount of net working capital in relation to sales, the better the current financial health. The net working capital also indicates the 'margin of safety' for meeting the maturing short-term liabilities (claims).

Gross Working Capital: Gross working capital is considered to be equal to the total current assets required by a business firm. In this context, the concept of total current assets has a broader application. It is intended to denote the total investment of funds in current assets for operation purposes or the total requirement of funds for current assets regardless of the financing sources.

Components of Gross Working Capital: The constituent parts of gross working capital, otherwise called current assets, are:

- Advance given for purchases of raw materials and stores, etc.
- Inventories:
    - Raw materials, stores and packing material, spare parts, etc.
    - Work-in-process.
    - Finished goods.
- Book debtors or credit to customers.
- Marketable investment (securities) on short-term basis:
    - To earn something on temporary surplus cash.
    - To meet the requirement of offering security for some facilities and/ or contract with government or some other agencies.
- Cash and bank balances.

Estimating Working Capital Requirements: As stated earlier both excess and shortage of working capital are harmful for the health of an enterprise. It is, therefore, essential to correctly assess the amount of working capital for an enterprise.

The following methods can be used to estimate the amount of working capital.

- **Operating Cycle Method:** If for instance, the operating cycle of an enterprise is four months, it means the cycle is repeated three times in a year. The amount of working capital required would be one-third (1/3) of the amount of annual operating expenses.
- **Assets and Liabilities Method:** The quantity of working capital can additionally be assessed based on the current assets necessary for the business and the credit options (current liabilities) accessible for obtaining current assets.

Institutional Support

1. **Small Industries Development Organization (SIDO):** Small Industries Development Organization (SIDO) operates as an office under the Department of SSI and Subsidiary and Rural Industries (ARI). It serves as the top organization and a central body that creates, coordinates, and oversees policies and programs aimed at promoting and developing small-scale industries.

   The Development Commissioner leads SIDO. He receives support from several managers and advisors in the creation and execution of different training and management initiatives, consulting, industrial training, various opportunities for small industry development, industrial zones, and more.

   Functions of SIDO

   The main functions of the SIDO are classified into:

   - Co-ordination,
   - Industrial development, and
   - Extension

   These activities are executed through a system of national organizations and associated entities set up for particular purposes. At present, SIDO functions through 27 offices, 31 Small Industries Service Institutes (SISI), 37 outreach centers, 3 product and process development facilities, and 4 production sites.

   All small-scale industries, excluding those managed by specialized boards and organizations such as Khadi and Village Industries (KVI), Coir Boards, Central Silk Board, etc, fall within the scope of SIDO.

   The main tasks of SIDO in all three functional categories are:

   Functions Relating to Co-ordination:

   - Create a national strategy for the growth of small industries,
   - Align policies and initiatives of different state governments,
   - To uphold appropriate communication with pertinent Central Ministries, Planning Commission, State Governments, Financial Institutions, etc. , and
   - To synchronize the development programs of industrial regions.

   Functions Relating to Industrial Development:

   - Reservation of products for small industry production,
   - Collect information on imported consumer goods and then encourage

industrial units to produce those goods by providing coordinated assistance;

- Provides necessary support for the development of auxiliary units and
- Encourages small industries to actively participate in the purchase program of state stores by providing them with the necessary instructions, marketing advice and assistance.

Function Relating to Extension:

- Offers technical services for the preparation of technical processes, production planning, selection of appropriate machinery, and preparation of factory layout and design
- Delivers consulting and training services to enhance the competitiveness of small industries
- Offers marketing support to small industries for the effective sale of their products and
- Aids small-scale industries with financial research and information

2. **The Small Industries Service Institutes (SISI'S):** Small Industry Service Institutes (SISIs) were set up in each state to offer guidance and training to small and aspiring entrepreneurs. The Industrial Management Training Division of the DC SSI Office (New Delhi) oversees the activities of the SIS. A total of 28 SISIs and 30 branch SISIs have been established in state capitals and other locations throughout the country.

   SISI has a broad array of technological, managerial, and administrative responsibilities to fulfill.

   Functions of SISI

   - Help current and future entrepreneurs by providing technical and managerial advice such as assistance in choosing suitable machines and equipment, adopting recognized testing and quality standards, etc.
   - Implementation of EDP throughout the country;
   - Advise Central and State Governments on policy matters related to the development of small-scale industries;
   - Assists in testing, inspection and quality control of SSI raw materials and products;
   - Transmission of marketing data to SISI;
   - Recommend financial assistance to SSI from financial institutions;
   - Involvement of entrepreneurs in the purchase program of state stores;
   - Conduct financial and technical studies and prepare techno-economically viable reports on selected areas and branches of the economy.

3. **National Small Industries Corporation LTD. (NSIC):** National Small Industries Corporation Ltd. (NSIC), a company certified under ISO 9000, has been striving to achieve its goal of supporting, aiding, and fostering the

development of small-scale industries and small-scale services related to industries/companies in the country since it was established in 1955.

Functions of NSIC: NSIC offers an extensive assortment of services, primarily of a promotional nature, to small-scale industries.

Its main functions are to:

- Offer machines on hire purchase to small scale industries.
- Offer to rent equipment.
- Assistance in export marketing of small industrial products.
- Participate in the government bulk purchase program.
- Develop prototypes of machinery and equipment to be transferred to small-scale industries for commercial production.
- Distribution of basic raw materials among small industries through raw material warehouses.
- Assistance in the development and modernization of technology and the implementation of modernization programs for small industries.
- Organize trainings for various industries.
- Establishment of small-scale industry in other developing countries on a turnkey basis.
- Complete the construction of industrial zones.

4. **Small Industries Development Corporation (SIDCO):** Small Industries Development Corporation was created to support small-scale industries across many states. They engage in various activities to enhance small industries. These organizations assist Small Scale Industries (SSI) in multiple aspects, from installation to the commencement of production. In summary, they offer infrastructural support for small scale industries. Numerous underdeveloped regions have been improved with the assistance of SIDCO in the majority of states. Therefore, SIDCO also played a role in the distribution of industrial activities in various countries.

   Functions of SIDCO

   - **SIDCO supplies scarce raw materials:** Some of the limited raw materials are sourced either from the local market or from overseas and are provided to small industries that require them. For this purpose, SIDCO has multiple raw material storage facilities and these facilities deliver various limited raw materials according to the needs of the small-scale industries within the state.
   - **SIDCO provides marketing assistance:** The firm has introduced multiple initiatives to aid in the effective marketing of small enterprises. Indeed, the firm is involved in the procurement processes of state ministries as well as DGS and D (Directorate General of Supply and Disposal). SIDCO provides upfront payments for the collection and distribution of orders among different small-scale businesses. Additionally, SIDCO regularly arranges

buyer-seller meetings.

- **SIDCO assists in Bills discounting:** When minor units supply goods to ministries, the payment receipt is postponed. In this scenario, SIDCO reduces the invoices of the ministries, and up to 80% of the invoice amount is allocated to the supplier. This will assist the SSI units in addressing their working capital challenges.
- **SIDCO provides Export marketing assistance:** To enhance the export marketing for small-scale industries, SIDCO has created web pages, which allow for the presentation of small-scale industrial products in overseas markets and the acquisition of export orders. After the export order is placed, the Joint Export Manager of SIDCO organizes the provision of various services to facilitate the product's export. SIDCO also supports small-scale units that take part in international exhibitions.
- **SIDCO set up Captive power plants:** In order to ensure consistent and high-quality power, SIDCO has initiated the planning of its power plant for extensive industrial zones. There are currently proposals to establish these factories in 10 industrial zones.
- **SIDCO promotes skill development centers:** In order to supply skilled labor for different small-scale industries, skill centers are established in numerous industrial zones to educate workers in various industrial tasks and teach them to obtain contemporary skills.
- **SIDCO promotes women entrepreneurs:** A separate Women's Industrial Zone has been established at Tirumullaivoyal near Chennai to train women entrepreneurs in various small-scale industries.

Besides SIDCO, numerous organizations such as Tamil Nadu Small Industries Promotion Corporation (SIPCOT), Tamil Nadu Small Industries Corporation (TANSI), Tamil Nadu Industrial and Technical Consultancy Organization (ITCOT), and Tamil Nadu Industries Investment Corporation (TIIC) are assisting in the promotion of small-scale industries.

5. **District Industries Center (DIC):** The district industrial center is an organization established at the district level, aimed at facilitating the development of small and village industries within the region. Prior to the establishment of a DIC, a prospective entrepreneur needs to contact multiple agencies, many of which are situated far from their locality, to secure the required assistance and resources. This resulted in significant delays, and the inefficient use of time and money.

   Currently, different departments of the State Government have assigned relevant powers to the District Industrial Centre. Consequently, the entrepreneur receives all the essential support from one of his own agencies, or DIC.

   Functions of the District Industries Center

   - **SURVEY AND INVESTIGATION:** The regional industrial hub outlines current traditional and emerging industries along with available raw materials and workforce. It generates market predictions for different products. Additionally, it compiles technical-economic viability reports to offer investment guidance to businesspeople.

- **Training Courses:** The Regional Industry Center also arranges training sessions for entrepreneurs of small and micro enterprises. It serves as a link between entrepreneurs and small-scale industrial service agencies to present new and enhanced product lines and quality established by the latter.
- **Machinery and Equipment:** The industrial hub of the area provides information on where to buy machines and equipment and organizes the delivery of the machines for a nominal fee.
- **Raw Materials:** The Regional Industrial Center gathers details regarding the materials required by different units and arranges for their bulk purchase. This way, smaller units can acquire raw materials at an affordable cost.
- **Arrangements for Loans:** It establishes essential agreements with the primary banks and other financial entities to offer financial support to entrepreneurs. It additionally assesses the application and monitors the distribution of industrial credit in the area.
- **marketing:** District Industries Center performs market research and market development initiatives. He additionally arranges market locations, engages with public procurement officials, and updates entrepreneurs on market data.
- **Khadi and Village Industries (KVI):**The industrial hubs of the district focus on fostering the growth of khadi, village industries, and various cottage industries. It also works in close partnership with the State Khadi Board and facilitates training sessions for rural artisans.

7. **Small Industries Development Bank Of India (SIDBI):** SIDBI is a financial development organization functioning in India. It was established on April 2, 1990, and is based in Lucknow, with branches throughout the nation. It aims to offer refinancing solutions and short-term financing to industries and become the leading financial institution for the Micro, Small, and Medium Enterprises (MSME) sector. SIDBI also oversees the operations of institutions involved in related activities. SIDBI works under the Ministry of Financial Services, Government of India.

   Main Schemes of SIDBI

   - **National Equity Fund Scheme:** It provides capital support to small entrepreneurs starting projects in the micro sector.
   - **Technology Development & Modernization Fund Scheme:** It provides financial support to existing SSI units for technology up gradation and modernization.
   - **Single Window Scheme:** It offers both a term loan for fixed assets and a loan for working capital through the same agency.
   - **Composite Loan Scheme:** It is for equipment and/or working capital and

also for work sheds to artisans, village and cottage industries in Tiny Sector.

- **Mahila Udyam Nidhi (MUN) Scheme:** It offers equity assistance to female entrepreneurs for establishing initiatives in the Tiny Sector.
- **Scheme for financing activities relating to marketing of SSI products**: It offers support in different marketing-related tasks including marketing analysis, research and innovation, product enhancement, involvement in trade shows and exhibitions, brand advertising, and the creation of distribution channels such as showrooms, retail outlets, fitting rooms, etc.
- **Equipment Finance Scheme:** It is for the procurement of machinery/equipment, including Diesel Generator Sets, that are not associated with any particular project.
- **Venture Capital Scheme:** It seeks to motivate SSI companies/subcontractors to obtain capital equipment and technology for enhancing export capability/import replacement, encompassing the overall expenses of quality management and ISO-9000 certification and skill development.
- **ISO 9000 Scheme:** To address the expenses associated with consultation, documentation, audit, certification fees, along with the equipment and calibration tools necessary for acquiring ISO 9000 certification.
- **Micro Credit Scheme:** Meets the criteria for voluntary organizations that have been active for a minimum of 5 years; you possess valuable experience and a well-established network along with expertise in small savings through credit schemes associated with Self Help Groups (SHGs).

New Schemes

- To improve the export potential of SSI units
- Program for Marketing Support
- Infrastructure Improvement Program
- Program for obtaining ISO 9000 certification
- Factoring Services and
- Bills Re-discounting Program for domestic supply bills of SSIs

Major Schemes

1. **Technology Development & Modernization Fund:** SIDBI has launched a Technology Development and Modernization Fund (TDMF) scheme aimed at providing direct support to small-scale retail industries in order to motivate existing industrial units in the sector to upgrade their production facilities and implement enhanced and modern technology to improve their export potential. Assistance through the program can encompass investment expenses arising from the purchase of assets, acquisition of technical expertise, advancements in process technology and product quality, enhancements in packaging, as well as expenses related to total quality management (TQM) and obtaining the ISO 9000 series certification.

In July 1996, SIDBI empowered SFCs and promotional banks to provide loans for modernization initiatives costing up to Rs. 50 lakhs. The TDMF scheme was expanded in scope effective from 09. 01. 1997. Non-exporting units and units within the SSI sector are now qualified for subsidies under this program.

2. **National Equity Fund:** Small Industries Development Bank of India (SIDBI) National Equity Fund (NEF) offers equity-like support to SSI units, small units with a service charge of five percent. The scope of this scheme was broadened in 2000-01 by raising the loan limit from Rs. 6. 25 lakhs to Rs. 10 lakhs and a project cost cap from Rs. 25 lakhs to Rs. 50 lakhs.

    a) The following are qualified for assistance under the scheme: New projects in small and small industries for manufacturing products in conservation or processing, regardless of their location (excluding units in the capital region). The aforementioned existing micro and small industrial units and service firms (including those that have previously received NEF assistance) that undertake expansion, modernization, technology upgrading, and diversification regardless of their location (excluding in metropolitan areas).

       Sick units within the small and small sector, including the previously mentioned service firms, which are deemed feasible regardless of the location of the units (excluding units in the capital region).

    b) For new projects from existing units and service companies, the project cost (including labor margins) cannot surpass 50,000 thousand rubles. Costs related to expansion/modernization/upgradation of technology, diversification, or renewal should not exceed Rs 50 lakh for each project.

    c) The manager's prior financial contribution, which is 10% of the project costs, will remain unchanged. However, the limit on the concessional loan subsidy within the scheme has been raised from the current 15 percent of Rs. 1000000 per project to 25 percent of the project cost.

    d) 30% of the investment is allocated for small units.

Khadi and Village Industries Commission (KVIC)

Before gaining independence, Mahatma Gandhi spearheaded the growth of Khadi and Village Industries, which were entirely self-sufficient from the government. Following independence, the Government of India was authorized to incorporate the growth of Khadi and Village Industries into the comprehensive framework of the Five-Year Plans. In reaction, the Government of India established the Khadi and Village Industries Commission (KVIC), a legal entity formed by an Act of Parliament. This organization was founded in 1956 with the goal of promoting and supporting Khadi and Village Industries to enhance the rural economy.

The Khadi and Village Industries program plays a significant role in creating job opportunities for rural artisans, particularly those from lower socio-economic backgrounds. As agriculture has diminished its capacity to produce supplementary employment for the continuously increasing workforce in rural regions, the Khadi and Village Industries Commission (KVIC) has become more significant in offering an alternative and sufficient income source for rural inhabitants.

As per RTI, the Khadi and Village Industries Commission (KVIC) is a legal entity formed by the Government of India in April 1957 under the "Khadi and Village Industries Commission Act, 1956. "It serves as a leading organization under the Ministry of Micro, Small and Medium Enterprises that seeks to plan, promote, strengthen, organize, and support the creation and progress of khadi and village industries in rural areas, collaborating with other rural development entities where necessary."

The Role of KVIC in Business

The Khadi and Village Industries Commission is a legally established body formed by the Government of India in April 1957 under the Khadi and Village Industries Commission Act of 1956. The Khadi and Village Industries Commission (KVIC) serves as an agency of the Government of India with a constitutional duty. It operates under the supervision of the Ministry of Micro, Small and Medium Enterprises.

Objectives of KVIC

KVIC's objectives include:

- Promoting Khadi in rural regions
- Providing work
- Producing saleable goods; and
- Fostering self-reliance among the underprivileged.

Functions of KVIC

- Creating a raw material reserve and putting it into action so that it may be allocated to producers
- Establishing shared service facilities for raw material processing, including semi-finished products
- Khadi and Village Industries items, as well as handicrafts, are being marketed and sold.
- Promoting research into manufacturing processes and equipment for the village industries sector
- Providing financial assistance to individuals and organisations for the development and operation of Khadi and village industries

National Bank for Agriculture and Rural Development (NABARD)

The National Bank for Agriculture and Rural Development, commonly referred to as NABARD, serves as the principal regulator for the rural banking system in the country.

It is the primary development finance institution of the nation, created and owned by the Government of India. The bank's purpose is to provide and oversee loans to rural regions, marking the initial step towards enhancing rural development in the country.

NABARD has been entrusted with several functions in the field of agriculture and economic development, including policy formulation, planning and operations. NABARD fulfills these responsibilities effectively and strives to promote and develop rural industries such as agriculture, cottage industries, other small-scale industries and rural crafts to improve infrastructure and employment opportunities for people living in those areas.

This financial institution was founded by the Government of India in accordance with the National Agricultural and Development Bank Act of 1981. In simple terms, the National Bank of Agriculture and Rural Development, or NABARD, is the foremost and most specialized bank for agricultural and rural development in the nation.

What is the role of NABARD?

The National Bank for Agriculture and Rural Development, commonly known as NABARD, serves as the main regulatory entity for agriculture and rural development. These are:

- The National Bank for Agriculture and Rural Development, or NABARD, offers investment and production loans for a variety of activities and projects aimed at enhancing rural development and prosperity. As this bank is the central financing source for all such development initiatives, it is responsible for ensuring sufficient funding and promoting the projects.
- NABARD is tasked with coordinating all economic activities in rural regions with institutions engaged in development projects. It must establish connections with all key institutions, including the Government of India, Reserve Bank of India, State Governments, and other major entities involved in ongoing agricultural or rural development efforts.
- NABARD's goal is to oversee, develop rehabilitation plans, reorganize credit institutions, and provide training for personnel, among other things, to enhance the loan distribution system's absorption capacity and to build a robust institution for this purpose.
- The State Bank offers refinancing to all financial institutions that fund NABARD or agriculture and rural development projects within the nation.
- When a bank has financed a rural development project or initiative, it is NABARD's duty to oversee and assess the project or initiative.
- NABARD provides guidance to all client institutions and trains all entities engaged in rural development or those intending to engage in rural development in the future.
- Alongside all the aforementioned responsibilities, the Land and Rural Development Bank oversees a range of natural resource management programs.
- NABARD also supports Self Help Groups (SHGs) through its SHG Bank

Linkage Project, which aims to enhance SHG activities in rural regions and represents a significant advancement in rural development.

Functions of NABARD

The National Bank for Agriculture and Rural Development performs four basic functions in order to fulfil its roles. The NABARD's four major functions are credit functions, financial functions, supervisory functions, and development functions.

- **Credit transactions:** The National Bank for Agriculture and Rural Development, commonly known as NABARD, serves as a key lender in rural regions by offering credit transactions. In conjunction with these functions, the Bank manages, oversees, and regulates the credit distribution in the rural sectors of the nation.
- **Financial Activities:** NABARD has numerous client banks and institutions that aid and support rural development initiatives. The National Bank for Agriculture and Rural Development, commonly known as NABARD, extends loans to various client banks and institutions like craft industries, food parks, processing units, artisans, etc. to facilitate economic activities. Supervisory Responsibilities: As mentioned previously, NABARD is the leading body in charge of agriculture and rural development. Hence, this authority is tasked with overseeing and regulating all development endeavors and initiatives. For this objective, NABARD carries out monitoring activities that necessitate it to examine all client banks, institutions, as well as credit and non-credit entities involved in rural development projects.
- **Development activities:** As you may already be aware, the primary goal of the National Bank for Agriculture and Rural Development, or NABARD, is to emphasize establishing sustainable farming and fostering rural growth. In pursuit of this aim, the bank provides development initiatives. NABARD's developmental responsibilities encompass aiding rural banks in creating development action plans.

The National Bank for Agriculture and Rural Development, commonly known as NABARD, effectively carries out all the subsequent tasks and functions and plays an essential role in the agricultural achievement and rural advancement of the nation.

12

Project Report and Project Appraisal

Project Report

A project report is a written document associated with every investment. It includes the details upon which the project was evaluated and deemed viable. It encompasses information regarding economic, technical, financial, management, and production elements. It assists the entrepreneur in understanding the interests and aids in securing loans from banks or financial institutions.

The project report provides comprehensive details on the necessary land and buildings, yearly production capacity, production methodology, machinery and equipment along with their costs and technical specifications, raw material needs, electricity and water requirements, labor needs, market costs of the project, production, financial analyses, and the financial viability of the project.

Contents of a Project Report

1. **General Information:** The project report needs to incorporate details regarding the industry relevant to the project. It ought to present information about the industry's historical background, present condition, challenges, and future outlook. It must include details about the product to be created and the rationale behind the product selection if the proposed organization is a manufacturing entity. It should indicate the demand for the product in local, national, and international markets. It should distinctly outline the business opportunities and provide explanations for initiating the business.
2. **Executive Summary:** The project report ought to specify the business goals and the strategies through which the company can attain success. The project report needs to provide a comprehensive overview of the company regarding capital, operations, and approaches to business operations and execution. It should outline the assumptions and risks linked with the company in a broader sense.
3. **Organization Summary:** The project's report must include the organizational structure and model of the unit. It should clarify if the foundation of the ownership right is a limited company, a corporation, or a joint stock company. Information on the driver's biography, encompassing financial stability, needs to be provided. The project report has to specify the name, address, age, and experience of the owners or promoters of the suggested business.

4. **Project Description:** A concise overview of the project must be provided and must include information regarding the following:
    - Geographic location of the site,
    - Requirements for raw materials,
    - Production targets,
    - Space needed for the work shed,
    - Requirements for power,
    - Requirements for fuel,
    - Water supply needs,
    - Employment needs for both skilled and unskilled labor,
    - Technology chosen for the project,
    - Process of production,
    - Forecasted production quantities, unit costs,
    - Required pollution management plants.
    - If the company focuses on services, it should specify the types of services available to customers. It should explain how the service will be delivered to customers.

A. Marketing Plan

The project report needs to explicitly outline the anticipated overall demand for the product. It should specify the selling price of the product in the market. The approaches employed to seize market opportunities must also be included. If after-sales service is provided, this too needs to be highlighted in the project. It should detail the distribution method of the product from the production unit to the market. The subsequent information should be included in the project report:

- Customer categories,
- Intended markets,
- Market characteristics,
- Segmentation of the market,
- Market outlook for the future,
- Sales goals,
- Project's marketing expenses,
- Share of the market for the intended business,
- Need for the product in regional, national, and international markets,
- It should identify prospective product users and the distribution methods to be employed for delivering the product.

Capital Structure and operating cost

The project report should outline the overall capital needs of the project. It should identify the financing sources; it should also state the amount of both owner and borrowed funds. The project needs to detail working capital necessities as well as the source of supply. The total cost estimate of the project should be divided into land, building construction and earthworks, equipment and machinery, assorted fixed assets, initial and operating costs, and working capital.

Management Plan

The project report must include the following.

- Experience in business of the promoters of the business,
- Information regarding the management team,
- Roles and responsibilities of the team members,
- Existing personnel requirements of the organization,
- Approaches to managing the business,
- Strategies for recruiting and training personnel,
- Programs and policies of the management.

Financial Aspects

To evaluate the business's profitability, the anticipated profit statement and balance sheet need to be included in the project report. It should display the projected sales revenue, production expenses, gross profit, and net profit expected from the proposed unit. Alongside the aforementioned, a projected balance sheet, cash flow statement, and cash flow statement must be prepared on an annual basis for at least 3-5 years.

The income statement and cash flow forecast should feature a three-year overview, with comprehensive monthly details for the first year and quarterly information for the second and third years. The project report must contain ROI and ROI. The accounting system and inventory control system utilized are typically examined in this section of the project report. The project report needs to indicate whether the company is economically and financially sustainable.

Technical Aspects

The project report offers details regarding the technology and technical features of the project. This encompasses details about the technology selected for the project, the manufacturing process, equipment capacity, pollution management systems, and so forth.

Project Implementation

Each suggested business unit is required to develop a project timetable. It must specify the period during which tasks associated with creating the company can take place. Implementation plans outline the schedule for project development and conclusion.

Social responsibility

The suggested units will obtain their support from the community. Thus, its input to the community regarding employment, income, exports, and infrastructure. The business results should be represented in the project report.

Project Appraisal

Project appraisal is a process in which the lending financial institution offers an unbiased and impartial evaluation of different components of an investment proposal to facilitate a financial decision. Project evaluation refers to the analysis of the project's economic, social, and financial viability. This entails a comprehensive review of the project.

Project evaluation can be described as a manager examining another essential and thorough project submitted by a project implementer who has minimal or no participation in its development and who can therefore adopt an independent, objective, and unemotional perspective on the project overall and its various elements.

The appraisal of proposed project includes the following analysis:

- **Economic analysis:** Economic analysis evaluates a project from the viewpoint of the overall economy, questioning if the benefits of the latter are as significant as the project's costs, which warrant the investment in it.
- **Financial analysis:** The aim of assessing the financial components of a project is typically to guarantee the financial circumstances of the project to secure seamless implementation and effective functioning.
- **Market analysis:** Financial organizations assess the initiative to confirm that the specifics of the investment are financially viable. He analyzes the marketing reach of the initiative and its worth to the national economy by evaluating consumption trends and possible demand for the project.

Market analysis covers the following:

- Anticipated market for the product
- Analysis of market opportunity and specifying objectives
- Planning the process of marketing the product
- Organization for the marketing process
- Life cycle of the product

B. Technical analysis

Technical evaluation of a project generally encompasses a thorough examination of the following:

a) **Location and site:** Many factors affect the location of an industry, as it can significantly affect production costs and distribution efficiency, operating environment, etc. Important factors affecting the location of the industry are raw materials, proximity to the market, availability of water, electricity, transportation options, labor and employment legislation, taxes, incentives, subsidies, etc. Factors to consider when choosing a location are carrying capacity, availability of water, sewage disposal, etc.

b) **Size of the plant/scale of operation**: The dimensions of the facility dictate the financial and financial obligations of the project. An important aspect of size is available process technology. Equipment is often standardized according to certain capacities of production sectors. Therefore, the working capacity of such sectors is only available at certain times.

c) **Technical Feasibility**: The appraisal of the technical aspects involves scrutiny of such aspects of the project as:
   - Technology selected
   - Technical collaboration arrangements made
   - Capacity/Size of the project
   - Selection of plant, machinery and equipment
   - Plant layout and factory building
   - Technical and engineering services

C. Organizational analysis

As lenders and development entities, banks and other financial institutions focus significantly on the necessity for project execution and effective organization alongside responsible management. In the assessment of the project, these two facets of the project are analyzed. If either area is not executed properly, immediate corrective measures are suggested for the business. The aim of this assessment is to guarantee the proper execution of the project. Different organizational elements comprise organization, structure, recruitment, training, and development, etc.

D. Managerial aspects

Even a good project can fail if the management is incompetent. It was rightly pointed out that if the project is weekly, it can be improved, but if the managers are weekly and lack business skills, it is difficult to turn the situation around.

To protect against this issue, financial institutions can oversee assisted entities. Financial institutions might designate board-appointed directors to the boards of any MRTP companies they aid. Companies function, and industries (development and regulation) enable the government to oversee management, including the takeover of management of industrial firms. All this illustrates the significance of having an effective management strategy to prevent poor management. If management factors are assessed properly from the outset, issues in this domain can be largely circumvented in the future.

Feasibility Analysis

a) **Market Analysis:** Market feasibility involves two elements: the overall demand for the suggested product/service and the market share of the initiative being examined. A variety of data and suitable forecasting techniques are necessary for this market evaluation. The information needed is:

- What would the total demand be for the suggested product or service?
- What would the market share of the project being evaluated be?
- Historical and present demand trends
- Historical and present supply situation
- Production capabilities and limitations
- Trade imports and exports
- Type of competition
- Cost structure
- Elasticity of demand
- Consumer behaviour:
  - Motivation,
  - Attitudes,
  - Preferences, requirements.
  - Distribution channels: marketing policies.

b) **Technical analysis:** The aim of the technical analysis is to determine whether the prerequisites for the effective execution of the project have been considered and if adequately sound decisions have been made regarding factors like location, size, etc. The key inquiries in technical analysis are:

- Have the necessary prerequisites for project success been evaluated?
- Appropriate selections concerning location, dimensions, processes, machinery, etc.
- Initial tests and research
- Accessibility of raw materials, energy, and other resources.
- Ideal operation sales.
- Selection of an appropriate production method
- Selection of suitable machinery and equipment
- Treatment of effluents and disposal of waste
- Effective arrangement of plant and structures
- Feasible working timetables
- Technology that is socially acceptable

c) **Financial analysis:** A financial analysis is necessary to determine if the proposed project is financially feasible in the sense that it can cover the burden of the maintenance department and if the proposed project meets the return expectations of the capital producers.

The following points should be considered during a financial assessment.

- Is the project financially feasible?
- Debt servicing
- Achieving return expectations

- Cost investment and phasing
- Funding methods
- Breakeven threshold
- Cash inflows in the project
- Is the investment worthwhile?
- Net present worth
- Internal return rate
- Repayment period
- Risk level

d) **Economic/Social Cost-benefit Analysis**: This involves evaluating the project from a broader social perspective, focusing on the social costs and benefits of the project, which can often differ from the costs and benefits of monitoring it. The social benefits analysis questions are as follows.
   - What are the immediate economic advantages and disadvantages of the project assessed using shadow (efficiency) prices rather than market prices?
   - What would be the effect of the project on the distribution of income within society?
   - What would be the effect of the project on the savings and investment levels in society?
   - What would be the project's contribution to achieving goals such as self-sufficiency, employment, and social order?

e) **Managerial analysis**
   - History of performance in previous project
   - Inventiveness of the promoter
   - Comprehension of business
   - Dedication to the project
   - Integrity

f) **Social analysis**
   - Impact of project on quality of:
     - Air
     - Water
     - Noise
     - Vegetation
     - Human life
   - Significant initiatives like these result in ecological harm
     - Power plants
     - Irrigation schemes
     - Industries like leather processing, chemicals etc.
   - Likely damage & the cost of restoration.

13

Structure and Government Policy Support

Structure of a business or legal framework

1. **Sole proprietorship:** A private company is a form where an individual is the exclusive proprietor of the business. It constitutes a straightforward setup where the owner is one person who manages the everyday operations of the company. In this case, the business is not regarded as an independent entity. Therefore, earnings and debts are merged into their personal accounts, which are assessed and taxed at the individual level.

   Advantages of the sole proprietorship structure include:

   - Complete Control: As a sole proprietor, you have the right to make all business decisions and do not have to negotiate with other partners, managers or shareholders as in other structures.
   - Easier Startup: The creation of a private company does not necessitate completing forms or remitting a state fee. This can ease the procedure and render it more affordable.
   - Simple tax reporting: Because a business is not legally a separate entity, your income tax return includes the expenses and income of the business. As an added bonus, you can use business losses to offset your earned income, which can result in a higher taxable income.
   - Privacy: Sole proprietors do not have to file annual reports with state or federal governments. Therefore, the company is not disclosed in the same way as in other structures.

2. **Partnership:** When two or more people own a business entity, it is called a partnership. They are similar to a sole proprietor, because the business does not have a separate legal recognition in the market. Instead, the company and owners are treated as a single entity.

   It is a business structure where each owner actively participates in the day-to-day operations of the business. They make all the decisions about the growth and expansion of the companies together. In addition, profits and debts are shared equally among all partners.

   Advantages of a partnership structure include:

   - **Simple start-up:** Forming a partnership does not require any paperwork for the federal government, although your state government may need to fill out some forms

- **Few Tax forms:** In cooperation, the company does not pay taxes on its income, so it does not have to submit corporate income statements. Instead, profits and losses are transferred to individual partners' income tax returns.
- **Shared Finances:** A co-owner can reduce the financial burden of starting a business, as partners can share purchasing and overhead costs. Banks are also more likely to offer loans to businesses with multiple owners, which can help with early-stage business financing.
- **Combined knowledge:** A partner can help grow your business by providing skills or expertise in areas you are unfamiliar with. They also bring a new perspective to important decisions such as business management.

3. **Limited Liability Company (LLC):** An LLC is a type of structure that combines features of both partnerships and corporations. Profits and losses are distributed equally to shareholders according to their investment share. Therefore, uniform taxation applies here. There are no restrictions on becoming a member of an LLC. Most countries allow individuals, corporations and even foreign entities to participate. Some states may have a single member LLC where one owner controls the entire business.

   Advantages and Disadvantages of LLCs: The main reason business owners choose to register their business as an LLC is to limit personal liability for themselves and their partners or investors. Many consider an LLC to be a mix of a partnership, which is a direct business agreement between two or more owners, and a corporation with some liability protection.

   While LLCs have some attractive features, they also have several disadvantages. Depending on state law, an LLC may be terminated by a member's death or bankruptcy. A business can exist forever.

4. **Private limited company:** A limited liability company is a type of legal entity intended for small businesses. This form of business structure restricts the owner's liability to their assets, caps the number of shareholders at 200, and governs the public business conduct of shareholders. Limited risk to personal assets - Shareholders of a private limited company enjoy limited liability. This indicates that you, as a shareholder, are required to cover the company's liabilities solely up to the level of your investment. Shareholders are not personally liable, meaning they are not required to fulfill the company's obligations using their personal finances.

   Advantages of a Private limited company structure includes:

   - **Legal Entity:** A PLC has a separate legal entity that is separate from you. This means that the Company is responsible for managing its assets and liabilities, debtors and creditors. You shall not be responsible for any losses suffered by the Company. As a result, creditors are unable to pursue you to reclaim the funds.
   - **Raising Capital:** Although establishing a limited liability company involves compliance obligations, business owners favor it since it assists in raising

capital through equity, growing, and simultaneously capping liability.

- **Trustworthiness:** Indian firms are registered with the ROC in accordance with the Companies Act, 2013. Anyone is able to verify company information via the MCA portal. Furthermore, details about all directors are given during the registration process. Consequently, the PLC-style business structure is considered more dependable.
- **Continue Existence:** A company possesses a "perpetual successor," signifying that it remains continuous or uninterrupted until it is legally dissolved. The company, as an independent legal entity, is unaffected by the death or resignation of a member, but continues to persist regardless of alterations in membership.

Institutional Support

Entrepreneurs need support from external parties who support the preparation and evaluation of projects. Support for technological improvement of business operations is also needed. Entrepreneurs need money for their initial investment. After a few years of business, entrepreneurs try to expand or diversify their business. Expansion or diversification requires large investments. They must be provided with investment support tools to expand and diversify their business.

That is why entrepreneurs need a financial support system.

Management assistance for small businesses is offered by the development institute, the management institute, financial institutions, and universities. To aid small enterprises in their marketing initiatives. The Government of India and state governments have set up various institutions such as,

- National Small Business Corporation
- State Small Business Corporation.
- Small Business Development Organization
- Khadi and Rural Industries Commission,
- Export Promotion Councils,
- Commodity Boards,
- State Trading Corporation, Minerals and Metals Trading Corporation, etc.,

To strengthen the marketing efforts of SSIs.

The government has broadened numerous forms of support initiatives, including market analysis, sharing information pertaining to the market and different elements of marketing, outsourcing exchanges, meetings between buyers and sellers, market consulting services, trade facilities, and export marketing support. Quality marketing program, collateral promotion, advertising, exhibitions, fairs, exhibitions, training programs, seminars, open house discussions, etc.

Industrial Policy Resolutions in India

Since gaining independence, the central government has released six industrial policy resolutions designed to foster industrial development and outline the approach of government intervention and support. The Industrial Policy Resolution of 1948, which established the foundation for basic and strategic industries, recognized that cottage and small-scale industries are especially effective for optimizing local resources and achieving "local self-sufficiency" in certain categories of industrial goods. The supportive policy for cottage, village, and small industries was formulated in 1956, when the government opted to implement measures to enhance the competitiveness of small and village industries. The 1956 resolution highlighted the potential contribution of the SSI sector in creating job opportunities, harnessing local skills and capital resources, and facilitating integration with the larger industrial sector.

The Industrial Policy Statement of 1977 highlighted the broader decentralization of cottage and small-scale industries in rural locations and minor urban centers. The idea of district industrial centers was also proposed to provide small industrial services under one roof. The Industrial Policy of 1980 played a significant role in the development of the sector concerning auxiliary operations and nuclear power plants, whereas the Declaration (1990) stressed initiatives to boost the SSI sector's contribution to total exports, job creation, and industry dissemination to rural regions.

The industrial policy measures introduced in 1991 primarily aimed at the enhancement and promotion of small, tiny, and village industries. Additionally, a novel SSI integrated infrastructure development scheme was initiated, which involves state authorities and financial organizations, as well as an active involvement of NGOs, which played a role in modifying investment thresholds, participation, and more. To safeguard its interests and encourage their swift growth, the government has periodically launched various support initiatives in line with its policy, including reservation policy, reevaluation of maximum investment figures, modernization, technological upgrading, marketing support, and others.

The highlights of the Small-Scale Industries Policy of August 1991 are as follows:

- The investment limit for small sectors has been raised to Rs. 5 lakhs from Rs. 2 lakhs. This limit is now increased to Rs. 25 lakhs.
- They are classified as small-scale industries, regardless of their geographical location.
- The Small Industry Development Organization (SIDO) is identified as the main institution that aids small industries and promotes exports.
- A center for export development will be created within SIDO to assist small units through its network of field officers, aiming to enhance the sector's export activities.
- A technology development unit (TDC) will be set up at SIDO, which could offer technological resources to enhance the quality and competitiveness of products from the small sector.

- The handloom sector, which represents 30% of the nation's total textile production, will be reorganized to consider local and regional demands. The government's approach will be to encourage the use of handlooms to sustain employment in rural areas and to enhance the living standards of handloom weavers.
- The National Small Industries Corporation (NSIC) will concentrate on marketing consumer goods under a unified brand, and connections will be formed between NSIC and SSIDC.
- The limitations of the State Equity Fund Scheme will be broadened to include projects up to Rs. 10 lakh for equity subsidies (up to 15 percent).
- The Single Window Loan Scheme has also been expanded to accommodate projects up to Rs. 20 lakh, with a working capital limit of Rs. 10 lakhs.
- Small units may obtain equity support reaching 24 percent of the total investment from medium and large industries, state-owned enterprises, public investments, or foreign investments.

In accordance with the socio-economic objectives of national development plans, development banks have either individually or jointly introduced several innovative promotional programs. Some significant programs consist of a low cost loan program, seed capital assistance, venture capital support, repayment plans, etc. To ensure that the government begins to focus more on the challenges facing the SSI sector, a new Ministry of Small-Scale Industries and Land and Rural Life was established. The industries were established on October 14, 1999 in the office of an independent minister of state.

Incentives And Subsidies

The term "incentive" includes concessions and rewards. "Subsidy" means a single fixed payment made by the Government to an industry. It is awarded to an industry considered important to the national interest. The term "bonus" refers to a bonus or financial advantage given by the government to an industry to help it compete with other entities in the country or in foreign markets. It is given against income. A bounty benefits a specific industry, while a subsidy is given in the interest of the country.

These subsidies and incentives offer the following advantages:

- They serve as a source of motivation that draws potential
- Entrepreneurs to engage in the manufacturing sector.
- They motivate entrepreneurs to establish industries in underdeveloped areas.
- Through the provision of subsidies and incentives, the Government can:
  - Foster industrial growth evenly across all areas,
  - Nurture more new entrepreneurs, which results in entrepreneurial growth,
  - Enhance the capacity of entrepreneurs to successfully contend with competition,
  - Lessen the overall challenges faced by small-scale entrepreneurs.

Incentive Schemes

Different incentive programs provided by Central and State Governments, along with Union Territories, to the entrepreneurs in India are as follows:

1. **Incentives for Development of Industries in Backward Areas:** As a component of efforts to guarantee equitable regional growth, the Government of India has periodically declared different incentives and amenities for firms located in specific underdeveloped areas/regions. The State Council recognized 247 regions (about 70% of the country's area) as backward and eligible for support. Many state governments have added to this list for state-level grants. Below is a brief outline of the subsidy program designed for establishment of industries in selected backward areas/districts:

   a) **Concessional finance:** All Indian financial institutions namely Industrial Finance Corporation of India and Industrial Credit and Investment Corporation of India, Industrial Development Bank of India provides financial support on preferential terms to both new and ongoing industrial initiatives within the expansion program, regardless of the project's expenses. The government has chosen 247 districts. The benefits offered by these financial institutions take the form of a reduced interest rate.

   b) **Capital Investment Subsidy**: Providing financial support for capital investment is called capital investment support. It is usually given as a direct subsidy of 10 to 20 percent of the capital invested in industrial units in areas designated as backward regions/districts. It is provided by the central government.

   The salient features of the scheme are given below:

   i. **Quantum of subsidy:** When the program was first revealed in 1971, 10 percent of the investment in fixed capital such as land, buildings, and plant and machinery was refunded as a direct grant limited to Rs. 15 laws. This was increased to 15 percent effective March 1, 1973. However, the maximum payment payable is limited to Rs. 15 lacs per industrial unit.

      After the division of backward districts into (A), (B) and (C) categories the subsidy will be: (A) 25% subject to a maximum of Rs. 25 lacs; (B) 15% subject to a maximum of Rs. 15 lacs (c) 10% subject to a maximum of Rs. 10 lacs.

   ii. **Eligibility:** An industrial unit other than a department which has made investments in land, buildings and equipment and machinery on or after March 1, 1973, and which is located in the said districts/areas, is eligible to apply for the subsidy. Existing expansion, modernization and diversification units can also apply for support.

   iii. **Procedure for Claiming Subsidy:** The State Governments/Union Territory Administrations have designated disbursing agencies to administer the Investment Support Scheme. State Finance Corporation and financial institutions like IDBI, IFCI and ICICI are some of the institutions selected to disburse grants under the scheme. Each industrial unit to be established in a certain area is registered with the Director of Industry to request investment support. Units desirous of receiving investment subsidy may approach the disbursing authorities who in turn make a recommendation to the State Level Committee appointed in each State/Union Territory after verification etc.

2. **Tax deduction to new industrial undertakings set up in backward States and Union Territories:** Under Section 80-IA of the Income-tax Act, 1961, a deduction is allowed for profits derived from a new industrial undertaking or shipping or hotel business in computing taxable income. A deduction under this section is allowed to companies of 30 percent of the profits of the assessment year applicable to earlier years in which a hotel begins operations or an industrial establishment begins production or a ship is first commissioned. and the nine assessment years immediately following the first assessment year. If the taxpayer is a cooperative, a similar deduction is allowed for the initial tax year and eleven subsequent years, a deduction of 25% is allowed for corporate liability. Similarly, in case of new hotels which are set up in hilly area or rural area or place of pilgrimage or any other place indicated by the government, the deduction is 50 percent of the profit.

   The Finance Bill proposes to encourage decentralization of industrialization to remote and industrially backward areas to make a significant contribution to promoting industrialization in highly industrialized states and union territories. It is proposed to allow deduction under section 80-IA to such new industrial establishment situated in a State or Union Territory specified in Schedule 8 to the Income-tax Act and commencing production or production after 1.4.1993. 100 percent of the profits of the first five assessment years starting from the previous assessment year in which the industrial undertaking started production or manufacturing. In the following assessment years, a deduction of 30 percent for corporations and 25 percent for companies is applied to the profits of those companies. The combined deduction with both enhanced and normal interest is restricted to twelve assessment years for cooperative societies and ten assessment years for other assessments according to the current regulations.

3. **Subsidized Consultancy Services to Small Entrepreneurs:** Small entrepreneurs intending to set up rural, cottage, small or micro units or expand/diversify/modernize existing units will get low cost advisory services from Technical Consultancy Organizations (TCOs) sponsored by All India and State Finance and Promotion Agencies and Banks. . They must pay only 20% of the fees charged by the TCO for tasks such as feasibility studies, project reports, market studies, pre-investment studies, diagnostic studies and special studies and seeking technical assistance from financial institutions. etc. An entrepreneur who has initiated a project in one location and intends to launch a distinctly different project in another location may also be regarded as eligible for an additional project. In any circumstance, the same enterprise will not receive assistance for over two projects.

4. **Subsidy for Market Studies:** New entrepreneurs (local or NRI) who are entering the country for the first time in medium and/or intermediate industries may obtain TCOs to carry out market research for their products at a lower cost. IFCI would fund the market research preparation fee payable to the TCO up to 75% of the cost or Rs. 15,000 whichever is less. Support will be provided only to TCOs with one or more State or All India Financial Institutions or Development Agencies joining as shareholders or board members. The entrepreneur has to pay only 25% of the research costs.
5. **Adoption of Indigenous Technology:** Project promoters for the commercial use of indigenous technology will receive a grant from IFCI to cover interest payments during the first three years of project operation, subject to a ceiling of Rs. 5 lakh per annum. In eligible cases, the total amount of subsidy can be up to Rs. 25 lakhs for five years. The subsidy will be returned to the company after it has made interest payments to IFCI on due dates. To get cheap support, the project should be established with IFCI loan support and based on original technology. The organization executing the project must possess authorization to utilize this technology from the appropriate institution, such as Government Laboratories, Public Sector Undertakings, Universities, or other government-sanctioned entities in India. The technology must be one that has not been launched on a commercial scale in the nation, and for this, a certificate needs to be secured from the relevant authority. The intended product must be dependent on technology and not merely accessories, and the project should align with a national priority as identified by the government periodically.
6. **Special Facilities for Import of Raw Materials:** In the import policy of 1978-79, a special concession was introduced for import of raw materials in the case of industrial units located in backward areas or graduates/graduates or military personnel belonging to scheduled castes/tribes. The maximum license value is Rs. 5 lakh for new or proposed small scale units instead of Rs. 3 lakhs. They are also entitled to preferential treatment for channeled programs.
7. **Transport Subsidy**: The transport subsidy system of 1971 includes the provision of transport subsidy to industrial units located in selected regions to the extent of 50% of the transport costs of raw materials brought to the selected regions and the transport costs of finished products exported to the selected regions. The scheme covers the state of Jammu and Kashmir, Himachal Pradesh, the hilly regions of Uttar Pradesh and the north-eastern region which includes the states of Assam, Meghalaya, Nagaland, Tripura and Arunachal Pradesh, Andaman and Nikobar Islands, Mizoram and the Union Territories, Lakshadweep.

   The subsidy will be disbursed on the transportation expenses between the chosen rail heads and the site of the industrial units in the aforementioned states/union territories.

The highlights of the scheme are:

- Industrial units located in the aforementioned areas receive transport support for raw materials brought to them and finished products exported from there.
- In the state of Jammu and Kashmir and the North Eastern region, no transport allowance is allowed for the inland movement of raw materials and finished goods.
- In case of Jammu and Kashmir, transport allowance is provided between Pathankot railway head and location of industrial unit or between location of industrial unit and Jammu, whichever is nearer.
- Transport allowance is provided between Siliguri and the location of the industrial unit in the North Eastern States excluding Jammu and Kashmir. While calculating the transportation cost, the cost from Siliguri to the nearest railway station to the industrial unit is considered for raw materials and finished products. If any other mode of transport is used, the costs are limited to the amount that the industrial unit could pay if it used the said mode of transport.
- Road freight charges are determined based on the freight charges decided by the State/Union Territory Government from time to time or the actual freight charges paid, whichever is lower.
- When determining transportation costs, loading or unloading costs and other handling costs are taken into account.
- All new industrial units located in designated areas can receive a transport subsidy of 50% of the transport costs of raw materials and finished products.
- Existing industrial units are also entitled to receive a 50% subsidy on additional transport costs of raw materials and finished products resulting from significant expansion or diversification undertaken by them after the commencement of the transport subsidy. • The transport subsidy covers 50% of the steel transport costs from Hindustan Steel Limited's Gauhati Stockyard to the location of the industrial unit in the North East region.
- State/Union Territories Industries Departments will determine the system of pre-registration of industrial units eligible for transport subsidy.

8. **Incentives Available to SSI Units in Backward/Rural Industries Project Areas:** Certain special facilities and incentives which have been provided to the backward districts are enumerated below:
   - 15% direct subsidy on share capital up to Rs. 15 lakhs is allowed for units established in backward areas.
   - Allotment of factory or factory premises for industrial areas/areas and industrialized colonies on easy terms.
   - Interest-free loans in lieu of VAT paid/remitted by interstate SSI units are available for a maximum period of 7 years provided the loan amount does not exceed 8% of the capital investment in a given year.

- Government finance companies offer loans for the acquisition of fixed assets of Rs. 30 lakhs for joint stock enterprises and registered cooperative societies, and up to Rs. 15 lakhs for others, with relaxed margins and interest rates, and the loans will be extended over a longer repayment and moratorium duration.
- The National/State Government, either directly or through its subsidiary Valtion Teollisuuden Kehitysyhdistys, invests in or selectively takes part in the preference shares of joint stock companies to facilitate the establishment of medium and large industrial units. The state government is also exploring joint ventures with the private sector.
- SSI units located in backward areas and other industries with a capital investment in plant and machinery of up to Rs. 1 lakh are exempt from specific taxation in certain States:
    - New blocks built in the urban areas are completely free of electricity tax for 7 years.
    - New apartments are exempt from property tax for 5 years.
    - Industrial units established within municipalities are exempt from foreign tax for capital equipment and construction materials for a maximum of 3 years from the date of regular registration.
- Provision of fundamental regulated raw materials to SSI units as a priority and on very generous terms.
- State Governments have established independent testing laboratories on behalf of Indian Standards Institution, Council of Export Inspection, Ministry of Defence, Government of India and various other government organizations for manufacturing quality industrial products.
- To provide essential and advanced joint facilities, a network of Industrial Development Centres, Heat Treatment Centers and Joint Workshops has been set up across the states to equip SSI equipment with modern technologies and production methods.

9. **Seed Capital Assistance:** The goal of the system is to create a new generation of entrepreneurs who have the necessary qualities for entrepreneurship, but whose financial resources are limited. To meet the venture capital requirements of entrepreneurs, it is planned to increase the assistance with a nominal service fee. The system is expected to promote greater decentralization of ownership and control of industrial enterprises. IDBI developed a system that gained popularity as seed capital programs. If the project comes in a non-damaged area, the project would not be eligible. Thus, the financier himself would make the entire contribution of the implementer. It is reduced to the amount of the subsidy if the project takes place in backward areas, for example (category A, B or C). The maximum amount punishable is Rs. 5 lacs per project subject to certain conditions.

The scheme is administered through notified SIDCs and SFCs. IDBI provides direct assistance under the scheme in exceptional cases. Projects backed by commercial banks can also receive seed capital support. However, entrepreneurs must submit their application through the SFC/SIDC operating in the region.

10. **Industrial Licensing: July 1991 Changes:** The process of liberalization was finalized with the announcement of a new industrial policy in July 1991. The primary aspects of this policy concerning industrial licenses can be outlined as follows:
    - All projects will see the elimination of industrial licenses, with the exception of a short list of 18 industries that pertain to security and strategic considerations, social factors, hazardous chemicals, significant environmental matters, and elite consumption.
    - In projects that require imported capital goods, automatic settlement occurs if the availability of foreign currency is ensured by foreign capital or if the CIF value of the required imported goods is less than 25% of the total value of machinery and equipment. Up to a maximum value of Rs. 2 million.
    - Staged production programs implemented administratively on a case-by-case basis cannot be applied to new projects. They still manage existing projects with such programs.
    - Existing units are equipped with new broadband to produce any product without additional investment.
    - License exemption applies to all significant extensions of existing units.
    - All existing registration systems will be abolished.
    - From now on, contractors are obliged to submit an information memorandum on new projects and major expansions.
    - The mandatory convertible clause no longer applies to final loans granted by financial institutions for new projects.
11. **Taxation Benefits:** The taxation benefits available to small scale industries are explained below:
    - **Tax Holiday:** New small-scale industries are not subject to income tax under section 80J of the Act on their profits up to 6% (7.5% for companies) of the total income of the units in the assessment year in which the units commenced production, provided the small-scale units followed the procedures set out in section 80J. This exemption is available for a maximum of 5 years from the beginning of production.
    - **Depreciation Allowance:** According to Section 32 of the Income Tax Act, small industries have the right to deduct the prescribed level of deductions from the depreciation account of machinery and equipment, land and buildings. Incase of small-scale industries, deduction from the actual cost of plant and machinery is allowed up to Rs. 20 lakhs. Depreciation is calculated

by the reducing balance system. Full depreciation is available annually regardless of the number of days the asset is actually used. Sometimes units are offered additional compensation called extra duty compensation. Any machinery or plant costing less than Rs. 750 can be fully amortized in the year it is first used.

- **Development Rebate:** For new equipment and machinery, with the exception of office equipment or small-scale road transport vehicles, which are used entirely for production purposes, the following amount is allowed as a development credit under § 33, in addition to the usual amount. depreciation
  - In case of plant and machinery, 35% of the actual cost if installed before April 1, 1970 and 25% of that cost if installed after March 31, 1970.
  - If the plant and machinery were installed after March 31, 1967, since it is a capital expenditure on research related to the business of the unit, product development is given in the amounts mentioned.
- **Rehabilitation Allowance**: This is granted to small scale units, under Section 33B, whose business has been disturbed by:

  Riot or civil disturbance

  Floods, hurricanes, cyclones, earthquakes or other natural disasters.

  Accidental fire or explosion

  Action by an enemy.
  - For a newly established, renovated or revived small unit, a deduction of rehabilitation money can be made, which corresponds to 60% of the unit deduction compensation amount.
- **Publication of Books:** Under Section 80 of the Act, a small business engaged in book publishing is entitled to claim a deduction of 20 percent of the profits and gains of such business. "Books" in this section do not include newspapers, magazines, periodicals, journals, pamphlets, pamphlets and other similar publications.
- **Tax Benefits for Amalgamation of Sick Units:** Disease in large or small industries is quite common in the country and has become a national problem that has caused much concern. The total number of units affected is estimated to be over Rs. 2000 crowns. Government policy aimed to promote the consolidation of struggling units, and incentives were revealed to attract healthy units to assume control of the troubled units for the sake of public welfare. Tax advantages can potentially be obtained when merging struggling units.

12. **Incentive to the small scale/ancillary units acquiring ISO 9000 certificate:** In March 1994, the Government of India launched a program to encourage small scale/subsidiary units to obtain ISO-9000 or equivalent certification. As per the prevailing norms, the scheme reimburses up to 75% of the cost of obtaining an ISO-9000 (or equivalent) certificate, subject to a maximum of

Rs. 75,000/- per case. The existing procedural standards regarding sending copies of various documents such as payment receipts/payment receipts etc. issued to certification bodies and consultants along with the claim have been revised and have now been simplified. According to the revised procedural rules, an accountant's certificate is required to prove the costs of application fee, assessment/inspection fee, annual fee/license fee, training, technical advice, etc. (excluding hotel and travel expenses) must be submitted in lieu of certified copies of various vouchers/payment checks for ISO-9000 certification. Furthermore, the application form has been made simpler. The revised standards significantly lessen the documentation efforts required from small/subsidiary units and accelerate the processing of compensation applications.

13. **Credit Link Capital Subsidy Scheme for Technology Upgradation of SSIs-Guidelines:** The scheme provides 12 percent retroactive capital subsidy for technological innovation projects of SSI units in certain products/sub-sectors. Small Industries Development Bank of India (SIDBI) serves as the central institution.

    - **Type of Unit to be covered**
        - Existing SSI unit registered with the State Directorate of Industries and updated with latest technology with or without extension.
        - New SSI units are registered with the State Directorate of Industries and set up their establishments only with relevant approved and proven technology duly approved by the Technical Advisory Committee.
    - **The Scheme would cover the following products/sub-sectors in the SSI:-**
        - Leather and Leather Products including Footwear and Garments
        - Food Processing
        - Information Technology (Hardware)
        - Drugs and Pharmaceuticals
        - Auto Parts and Components
        - Electronic Industry particularly relating to Design and Measuring
        - Glass and Ceramic items including Tiles
        - Dyes and Intermediates
        - Toys
        - Tyres
        - Hand Tools
        - Bicycle Parts, and
        - Foundries-Ferrous and Cast Iron.

The above list of products/sub-sectors will be expanded as the scheme progresses with the approval of the Technical Advisory Committee constituted under this scheme.

References

Abrams, R. M. (1993). The Successful Business Plan: Secrets and Strategies. Rev, ed. Grants Pass: Oasis Press, Mediford.

Agnete, Alsos, G., Ljunggren, E., & Pettersen, L. (2003). Farm-based entrepreneurs: whattriggers the start-up of new business activities? Journal of Small Business and Enterprise Development, 10,435-3.

Alan Gleeson, Managing Director, Palo Alto Software UK, http://bplans.typepad.com/blog/2007/04/delivering_a_wi.html

Altieri, M. A., Funes-Monzote, F. R., & Petersen, P. (2012). Agro-ecologically efficient agricultural systems for smallholder farmers: contributions to food sovereignty. *Agronomy for Sustainable Development*, 32,1-13.

Andersén, J. (2012). A resource-based taxonomy of manufacturing MSMEs. International *Journal of Entrepreneurial Behavior and Research*, 18, 98-122.

Bairwa, S. L., Kalia, A., Meena, L. K., Lakra, K., & Kushwaha, S. (2014b). Agribusiness management education: a review on employment opportunities. *International Journal of Scientific and Research Publications,* 4(2), 1-4.

Bairwa, S. L., Lakra, K., Kushwaha, S., Meena, L. K.,& Kumar, P. (2014a).Agribusiness management education: a review on employment opportunities.*International Journal of Scientific and Research Publications*, 4(3), 1-4.

Bansal, A. (2011). Agri-business in India–vision 2020. *International Journal of Multidisciplinary Research*, 1(2), 62-72.

Bernier, L.,& Hafsi, T. (2003). The changing nature of public entrepreneurship. USA, Midwest Political Science Association Conference.

Boehlje, M., Roucan-Kane, M.,& Bröring, S. (2011). Future agribusiness challenges: strategic uncertainty, innovation and structural change. *International Food and Agribusiness Management Review*, 14, 53-82.

Chandramouli, P., Meti, K. S., Hirevenkangoudar, L. V.,& Hanchinal, S. N. (2007). Comparative analysis of entrepreneurial behaviour of farmers in irrigated and dry land areas of Raichur district of Karnataka. *Karnataka journal of agricultural science*, 20(2), 320-2.

DeLind, L. B. (2002). Place, work, and civic agriculture: common fields for cultivation. *Agriculture and Human Values*, 19, 217-4.

Dollinger, M. J. (2003). Entrepreneurship–Strategies and Resources. Pearson International Edition, New Jersey.

Drucker, Peter, F. (1985). *Innovation and entrepreneurship*. Harper & Row, New York. pp.250. Economic Outlook 2021-22

Elliott, R. (2013). The taste for green: the possibilities and dynamics of status differentiation through 'green' consumption. Poetics, 41, 294-2.

Faisal, I. (2010). Institutionalization of agricultural knowledge: management system for digital marginalized rural farming community. *Proceedings of a symposium onInnovation and Sustainable Development in Agriculture and Food, Montpellier*. pp. 10-22. France, 28 June-1st July 2010. France.

Feagan, R. (2007). The place of food: mapping out the 'local' in local food systems. *Progress in Human Geography,* 31, 23-42.

Gray, C. (2002). Entrepreneurship, resistance to change and growth in small firms. *Journal of Small Business and Enterprise Development,* 9(1), 61-72.

Hanf, C.,& Müller, R. (1997). Schlüsselaktivitäten betrieblicher Anpassung: Informations beschaffung, Wissensakquisition, Erwerb von Fähigkeiten' In: Schriften der Gesellschaft für Wirtschaft- und Sozialwissenschaften des Landbaus e.V., 33, 207-8.

Haugen, M. S., & Vik, J. (2008). Farmers as entrepreneurs: the case of farm-based tourism. *International Journal of Entrepreneurship and Small Business*, 6, 321-6.

Hill, T. H. How to Differentiate Yourself from Your Competitors', American Salesman.http:// www.allbusiness.com/company-activities-management/operations- qualitycontrol / 6197318-1.html, accessed on 1 December 2007. The 10 Customer Service Trends for 2010, http://smallbiztrends.com/2009/12/customer- service-trends-2010.html

Ikerd, J. (2008). Sustainable capitalism: a matter of ethics and morality. *Problems of Sustainable Development,* 3, 13-22.

Ikerd, J. E. (2011). Local food: revolution and reality. *Journal of Agricultural and Food Information*, 12, 49-57.

Kumar, A., Kashyap, P., Prusty, A. K., & Panwar, A. S. Process of entrepreneurship development.

Land Matrix (2015) Agricultural drivers. Land Matrix Global Observatory. Available at: http:// www.landmatrix.org/en/get-the-idea/agricultural-drivers (Accessed 30 October 2018).

Lans, T., Seuneke, P.,& Klerkx, L. (2013). Agricultural entrepreneurship, pp. 44–49. Carayannis E G. (Ed.) Encyclopedia of Creativity, Invention, Innovation and Entrepreneurship. Springer, New York. Brockhaus, R. H. and Horwitz P S. 1986. The psychology of the entrepreneur. Sexton D L. and R.W. Smilor (eds.), The art and science of entrepreneurship. Ballinger publishing company, Cambridge, pp. 25-48.

Lauwere, C., de, Verhaar, K., & Drost, H. (2002). Het Mysterie van het Ondernemerschap, boeren en tuinders op zoek naar nieuwe wegen in een dynamische maatschappij('The Mystery of Entrepreneurship; Farmers looking for new pathways in a dynamic society, in Dutch with English summary'). Wageningen University and Research Centre.

Lingelbach, D. C., De, La., Vina, L.,& Asel, P. (2005). What's distinctive about growth-oriented entrepre- neurship in developing countries? Working paper no. 1, Center for Global Entrepreneurship, UTSA College of Business. University of Texas, San Antonio, Texas.

Lowson, R. H.(2002). Strategic Operations Management: The New Competitive Advantage. *Journal of General Management*, 28(1), 36–56.

Man, T. W. Y., Lau, T.,& Chan, K. F. (2002). The competitiveness of small and medium enterprises - A conceptualization with focus on entrepreneurial competences. *Journal of Business Venturing*, 17, 123-142.

McElwee, G. (2008). A taxonomy of entrepreneurial farmers. International Journal of Entrepreneurship and Small Business, 6, 465-8.

McElwee, G., & Smith, R. (2012). Classifying the strategic capability of farmers: a segmentation framework. International Journal of Entrepreneurial Venturing, 4,111-1.

McMichael, P. (2005). Global development and the corporate food regime. Research in RuralSociology and Development, 11, 269-3.

Meredith, J. R., & Mantel, S. J. (2009). Project Management: A Managerial Approach. 7th edition, John Wiley & Sons.

McElwee, G. (2008). A taxonomy of entrepreneurial farmers. *International Journal of Entrepreneurship and Small Business,* 6, 465-8.

McElwee, G.,& Smith, R. (2012). Classifying the strategic capability of farmers: a segmentation framework. *International Journal of Entrepreneurial Venturing*, 4, 111-1.

McMichael, P. (2005). Global development and the corporate food regime. *Research in Rural Sociology and Development*, 11, 269-3.

Mishra, A. H., El-Osta, H.,& Shaik, S. (2010). Succession decisions in U. S. Family farm business. *Journal of Agricultural and Resource Economics*, 35,133-2.

Mullins, J. W. 'Why Business Plans Don't Deliver, Entrepreneurship, http://sloanreview.mit.edu/executive-adviser/articles/2009/2/5121/why-business-plans-dont-deliver/, accessed on 22 June 2009.

Murphy and Schwartz, An Exchange of Values (sixth of a series) by Frank Demmier, Professor of Entrepreneurship at the Donald H Jones Center for Entrepreneurship at Carnegie Mellon University; http://www.andrew.cmu.edu/user/fd0n/6%20An%20Exchange%20of%20Values.htm

Nandram, S. S.,& Samson, K. J. (2000). Successful entrepreneurship: more a matter of character than of knowledge (in Dutch with English summary). Nyenrode Centre for Entrepreneurship, Breukelen, The Netherlands, 242.

Pandey, G. (2013). Agripreneurship education and development: need of the day. *Asian Resonance*, 2(4), 155-7.

Pettinger, T. (2016). Problems of agriculture-market failure. Economicshelp-org. Available at https://www.economicshelp.org/blog/4977/economics/problems-of-agriculturemarket-failure/EconomicsHelp.org. (Accessed 30 October 2018)

Pyysiäinen, J., Anderson, A., McElwee, G.,& Vesala, K. (2006). Developing the entrepreneurial skills of farmers: some myths explored. *International Journal of Entrepreneur Behavior and Research Information*, 12, 21-39.

Reddy, S. S., Ram, P. R., Sastry, T. V. N., & Devi, I. B. (2004). Agricultural Economics. Oxford & IBH Publishing Co. Pvt. Ltd., New Delhi.

Reich, Robert, B. (2002). *The future of success*. Vintage- Penguin Random House, New York. pp.304.

Requier-Desjardins, D., Boucher, F.,& Cerdan, C. (2003). Globalization, competitive advantages and the evolution of production systems: rural food processing and localized agri-food systems in Latin-American countries. *Entrepreneurship and Regional Development,* 15, 49-67.

Richards, S. T.,& Bulkley, S. (2007). Agricultural entrepreneurs: the first and the forgotten? Research Paper 07-01. *Center for Employment Policy Entrepreneur Series*. Hudson Institute, Washington, DC.

Sah, P., Sujan, D. K.,& Kashyap, S. K. (2009). Role of agripreneurship in the development of rural area, Paper presentation in ICARD at Banaras Hindu University, Varanasi – 221005.

Shailesh, K., Gyanendra, S.,& Yadav V K. (2013). Factors influencing entrepreneurial behaviour of vegetable growers. *Indian Research Journal of Extension Education*, 13(1),16-19.

Singh, A. P. (2013). Strategies for developing agripreneurship among farming community in Uttar Pradesh, India. *Academicia: An International Multidisciplinary Research Journal*, 3(11), 1-12.

Singh, M. P. (2014). Entrepreneur and economic development: a study of role of various forms of entrepreneurs in economic development. *Global Journal of Multidisciplinary Studies*, 3(5). GJMS, Jabalpur MP, India.

Srinivas, I. A. S. B., Shekara, P. C., Murthy, L.,& Sahare, J. (2014). Agri-Clinics and Agri-Business Centres Success Stories of Agripreneurs. National Institute of Agricultural Extension Management (An Organization of Ministry of Agriculture, Govt. of India). Rajendranagar, Hyderabad-500030.

Stevenson, H., Roberts, M.,& Bhide, A. (1999). The entrepreneurial venture, Sahlman W (Ed.). Harvard Business School, US.

Stupnytska, A., Koch, K., MacBeath, A., Lawson, S.,& Matsui, K. (2014). Giving credit where it is due: how closing the credit gap for women-owned SMEs can drive global growth. Goldman Sachs Global Market Institute, New York.

Sullivan, P. (2017). Agri-entrepreneurship and their characterisitics. Enabling Agrientrepreneurship and Innovation: Empirical Evidence and Solutions for Conflict Regions and Transitioning Economies, 1-17. CABI International.

Trienekens, J. H. (2011). Agricultural value chains in developing countries: a framework for analysis. *International Food and Agribusiness Management Review*, 14, 51-83.

UNIDO. (2015). Agribusiness and rural entrepreneurship development. United Nations Industrial Development Organization, New York.

University of Western Cape, School of Business and Finance, MAN620/ADM720 Project Management, Lecture Session 1: 26th January 2010

Uplaonkar, S. S.,&Biradar.(2015). Development of agriculture in India through agripreneurs. *International Journal of Applied Research*, 1(9), 1063-66.

Vahoniya, D. R., Halpati, J., Baruri, S., & Garval, A. (2022). Start-up India: Eligibility, Benefit and current scenario. The Pharma Innovation, 11(6), 2630-2636.

Vahoniya, D. R., Baruri, S., Halpati, J., & Garval, A. (2022). Start up India: Agriculture and allied sector scheme. International Journal of All Research Education and Scientific Methods, 10 (7), 1770-1789.

Verhees, F., Lans, T.,& Verstegen, J. (2012). The influence of market- and entrepreneurial orientation on strategic marketing choices: the cases of Dutch farmers and horticultural growers. *Journal on Chain and Network Science*, 12(2), 167-180.

Verma, M., Arya, L., Kashyap, P., & Tyagi, R. Scope and importance of agri-entrepreneurship.

Vogel, S. J. (2012). Multi-enterprising farm households: the importance of their alternative business ventures in the rural economy. USDA-ERS Economic Information Bulletin 101. USDA Economic Research Service, Washington, DC.

Ward, S. "Thinking of Starting a Small Business? Part 4: Avoiding Business Failure', About com Guide, http://sbinfocanada.about.com/cs/startup/a/startownbiz_4.htm Bareham, S.Planning to Fail Can Spell Success. http://www.canadaone.com/ezine/sept99/business_success.html 'Some of the Reasons Why Business Fail and How to Avoid Them, Entrepreneur Weekly. Issue 36, 3 October 1996

Yukl, Gary, A. (1994). Leadership in organizations. Prentice Hall, New York. pp.498.

Web references

http://cpa.utk.edu/pdffiles/adc24.pdf

http://dcmsme.gov.in/howtosetup/finance.htm, Development commissioner MSME, INDIA on 8/8/2018, 4.59 Pm

http://depts.washington.edu/pmgroup/GlobalCongress2006/PDS01_wp.pdf - Triple Contraint. www.method123.com/affiliates/method123-ebook.pdf - Phases of Project Management

http://entrepreneurship.mit.edu/15975/resources/Guidebook.pdf

http://fiddlerontheproject.wdfiles.com/local--files/start/thebalancingact.pdf

http://sbinfocanada.about.com/od/businessplans/a/execsumexample http://searchcio.techtarget.com/definition/project-management

http://www.asiaentrepreneurshipjournal.com/AJESIII2Okpara.pdf

http://www.atento.com/content/our_commitment_with_customers.mimp

http://www.brazzil.com/2004/html/articles/aug04/p121aug04.htm Headd. B. Business Success:Factors Leading to Surviving and Closing Successfully, Center for Economic Studies, U.S. Bureau of the Census, Working Paper #CES-WP-01-01, January 2001;Advocacyfundedresearch by Richard J. Boden (Research Summary #204)

http://www.businessdictionary.com/definition/market-demand.html

http://www.ecs-group.com/uk/m12445-e12189-fostering.sense,commitment.html

http://www.genco.com/Logistics-Articles/article.php?aid=800900305

http://www.ibbi.gov.in/

http://www.independent-consulting-bootcamp.com/difference-between-a-project-and-a-program.html

http://www.pmi.org/About-Us/About-Us-What-is-Project-Management.aspx

https://101entrepreneurship.org/agricultural-entrepreneurship-agripreneur/

https://accountlearning.com/project-appraisal-meaning-necessity-four-major-criteria/ accessed on 12/07/2018

https://accountlearning.com/project-report-meaning-contents-project-report/accessedon 12/07/2018

https://aim.gov.in/what-is-an-incubator.php

https://books.google.co.in/books?id=shfi8945CrIC&pg=PA8&lpg=PA8&dq=Increasingly+complex+and+technical+products&source=bl&ots=lAAxp5YCxO&sig=G8wwc0RZqEZDd6PGbfg6ZDuLkRI&hl=en&sa=X&ved=0ahUKEwi66dmV-qbLAhUDG44KHeY1CesQ6AEIIjAB#v=onepage&q=Increasingly%20complex%20and%20 technical%20products&f=false

https://bookshelf.vitalsource.com/#/books/9781119793120

https://ebooks.inflibnet.ac.in/hsp15/chapter/creativity-and-innovation/

https://keydifferences.com/difference-between-pert-and-cpm.html

https://openstax.org/books/entrepreneurship/pages/4-2-creativity-innovation-and-inventionhow-they-differ

https://pmstudycircle.com/2012/08/what-is-a-functional-organization-structure/

https://startup.gujarat.gov.in/home

https://www.businessmanagementideas.com/essays/agripreneurship/essay-on-agripreneurshipindia-business/17489

https://www.educationalstuffs.in/cpm-and-pert-1/

https://www.educationalstuffs.in/cpm-and-pert-1/

https://www.educationalstuffs.in/cpm-and-pert-1/

https://www.google.co.in/search?biw=1366&bih=599&noj=1&q=interpersonal+skills+meaning&oq= https://www.google.co.in/search?biw=1366&bih=599&noj=1&q=Problem-solving+skills+meaning&oq=Problem-solving+skills+meaning&gs_l=serp.3..0i30j0i5i30.4262.5571.0.6138.2.2.0.0.0.0.144.276.0j2.2.0....0...1c.1.64.serp..0.2.276.LkkQHv82APc

https://www.google.co.in search ? biw = 1366 &bih=599&noj=1 & q=communication+ skills +meaning &oq=

https://www.igi-global.com/dictionary/technological-advance/53595

https://www.pmi.org/learning/library/transforming-functionally-oriented-project-driven-3741

https://www.projectmanagementlearning.com/what-are-the-advantages-and-disadvantages-ofa-functional-organizational-structure.html

https://www.projectmanagementlearning.com/what-is-the-difference-between-project-drivenand-non-project-driven-organizations.html

https://www.researchgate.net/publication/46510531_The_Role_of_Creativity_in_Entrepreneurship

https://www.simplilearn.com/what-is-project-management-information-system-article

https://www.slideshare.net/kunaldrizzy/creativity-and-innovation-in-entrepreneurship

https://www.startupindia.gov.in

https://www.startupindia.gov.in

https://www.startupindia.gov.in/

https://www.startupindia.gov.in/content/dam/investindia/Templates/public/notification/Relaxed_Norms_of_Public_Procurement_for_Startups/1.%20Relaxed_Norms_of_Public_Procurement_for_Startups.pdf.

https://www.startupindia.gov.in/content/sih/en/Startupgov/startup-recognition-page.html.

http://www.startupsmart.com.au/planning/business-planning/2010-10-13/low-risk-planning-on-the-run-for-start-ups.html

https://www.youtube.com/watch?v=FXJUDyqobbM

https://www.youtube.com/watch?v=HD-yFYQ5UPM

https://www.youtube.com/watch?v=zO2LdDpx-Tc

